W9-BVZ-248

THE WORST OF TIMES

THE WORST OF TIMES

HOW LIFE ON EARTH SURVIVED EIGHTY MILLION YEARS OF EXTINCTIONS

PAUL B. WIGNALL

PRINCETON UNIVERSITY PRESS
Princeton and Oxford

Published by Princeton University Press, 41 William Street, Princeton,
New Jersey 08540
In the United Kingdom: Princeton University Press, 6 Oxford Street,
Woodstock, Oxfordshire OX20 1TW
press.princeton.edu
Jacket art © Dai Mar Tamarack/Shutterstock.
ISBN 978-0-691-14209-8
British Library Cataloging-in-Publication Data is available
This book has been composed in Sabon Next LT Pro &
Grotesque MT STD.
Printed on acid-free paper. ∞
Printed in the United States of America
10 9 8 7 6 5 4 3 2 1

FOR KAREN

CONTENTS

ILLUSTRATIONS   ix
ACKNOWLEDGMENTS   xi
PROLOGUE   xv

CHAPTER 1
A TIME OF DYING   1

CHAPTER 2
EXTINCTION IN THE SHADOWS   12

CHAPTER 3
THE KILLING SEAS   39

CHAPTER 4
TROUBLED TIMES IN THE TRIASSIC   89

CHAPTER 5
TRIASSIC DOWNFALL   117

CHAPTER 6
PANGEA'S FINAL BLOW   137

CHAPTER 7
PANGEA'S DEATH AND THE RISE OF RESILIENCE   154

NOTES   177
REFERENCES   179
INDEX   191

ILLUSTRATIONS

FIGURES

1.1. Subdivisions of geological time during the time of the Pangean supercontinent.   4

1.2. World map during the Permian Period, 260 million years ago.   7

2.1. Extinction rates since the start of the Cambrian Period, 540 million years ago, to the present day.   15

2.2. Fusulinid foram: A giant single-celled organism around 2 millimeters across.   16

3.1. Summary of carbon isotope changes and extinction levels during the Permo-Triassic mass extinction compared with the sedimentary beds seen at Meishan, China.   53

3.2. Two framboids of pyrite.   57

3.3. Sources of gas erupted during flood basalt volcanism.   65

3.4. Flow chart showing the chain of events responsible for the Permo-Triassic mass extinction.   69

3.5. Deoxygenation of ocean waters.   78

5.1. World map at the time of the Triassic-Jurassic boundary, 200 million years ago.   118

5.2. Triassic-Jurassic boundary rocks in southwestern England.   126

6.1. Cause and effect chain for Toarcian (Early Jurassic Period) extinction crisis.   151

7.1. The carbon pump.   171

PLATES

Plates follow page 108.

1. Bedding plane covered in Permian brachiopods, Spitsbergen, Norway.
2. David Bond pointing his rifle at the point where Capitanian brachiopods go extinct in Spitsbergen, Norway.
3. Simon Bottrell admiring some pillow basalt lavas of the Emeishan large igneous province, Yunnan, China.
4. The A team enjoying fieldwork in western Yunnan, China.
5. The Permian-Triassic boundary seen in Anatolia, Turkey.
6. Sausserberget mountain top in central Spitsbergen, Norway.
7. Early Triassic bivalves.
8. The Early Triassic world.
9. A conodont microfossil from the Triassic, China.
10. Quarry face in a Late Triassic reef, Bavaria.
11. A mountain of Triassic rock seen in the Dolomites of Northern Italy.
12. Rob Newton examining beautifully laminated shales of the Early Jurassic Period, Somerset coast, England.
13. Block of sandstone from the Late Triassic Period, Somerset coast, England.
14. Kettleness cliff, Yorkshire coast, England.
15. Bullet-shaped belemnites and an ammonite seen in Toarcian shales, Yorkshire, England.
16. Flood basalt landscape seen at Gásadalur in the Faroe Islands.

ACKNOWLEDGMENTS

This volume is the product of a quarter century spent trying to unravel what happened during the greatest series of catastrophes in the history of life. I originally started looking at the worst of them all, the Permo-Triassic mass extinction, because its sheer scale was awe inspiring but also, more prosaically, because the most famous mass extinction of the day, the one that killed the dinosaurs, was being intensively investigated. It would have been hard for a newly fledged geologist like myself to muscle into an already crowded field. Gradually the Permo-Triassic mass extinction has become the go-to disaster for extinction workers, and I have found myself drawn to study other examples, some of which had not even been discovered twenty-five years ago. Each is unique but they are all fascinating because they all share many attributes; it began to seem that when the world goes to hell, it does so in very similar ways. And so was born the idea for this book. Mass extinctions require a certain set of ingredients and when everything is just right, in an evil Goldilocks kind of way, a disaster will surely follow.

My initial research was in the company of Tony Hallam, my former PhD supervisor and long-term friend and collaborator. Even at that stage Tony already knew a thing or two about mass extinctions, having contributed to the debates about the downfall of the dinosaurs at the end of the Cretaceous. He had also single-handedly pioneered research on an earlier mass extinction 200 million years ago, at the end of the Triassic, a crisis that has only become trendy within the last decade. Our initial collaborations required us to visit the far-flung locations where Triassic rocks can be seen lying on Permian ones.

China turned out to be a key country for studying many of the mass extinctions discussed in this book, and right from the start Tony and I were fortunate to get expert guidance from Yin Hongfu and his student Lai Xulong of the China University of Geosciences in Wuhan. Hongfu has since retired but continues to produce influential research, as does Xulong, who has risen to a high position in Wuhan University and built a large research school. His research students have included Sun Yadong, Song Haijun, Jiang Haishui, Wang Lina, and Luo Genming, who have all become my valued and influential collaborators.

Having been appointed at Leeds University, I was able to build my own teams of extinction hunters including (in chronological order, oldest first) the following PhD students—Richard Twitchett, Rob Newton, David Bond, Eleanor John, James Witts, and Luke Faggetter—together with that rare breed of scientist, the postdoctoral research fellow: Stéphanie Védrine, Alex Dunhill, and Juan Carlos da Silva. I have enjoyed being able to share the many highs and lows of research with them all while of course allowing them do a lot of the hard work.

I would also like to thank colleagues and friends from around the world, some sadly no longer with us, who have shared my fascination with geological calamities and provided ideas, conversations, and different perspectives and have generally enlivened the already enjoyable life of a geologist: Thierry Adatte, Derek Ager, Tom Algeo, Jason Ali, Mike Benton, Dave Bottjer, Pat Brenchley, Mike Brookfield, Matthew Clapham, Jacopo dal Corso, Steve Grasby, Janos Haas, Steve Hesselbo, Jason Hilton, Yukio Isozaki, Dougal Jerram, Michael Joachmiski, Gerta Keller, Heinz Kozur, Kiyoko

Kuwahara, Cris Little, Cindy Looy, John McArthur, Jenny McElwain, Jared Morrow, Jonathan Payne, Jeff Peakall, Nereo Preto, Sara Pruss, Greg Racki, Sylvain Richoz, Alastair Ruffell, Ivan Savov, Dolf Seilacher, Mark Sephton, Jack Sepkoski, Sha Jingeng, Mike Simms, Bas van de Schootbrugge, Wang Wei, Mike Widdowson, and Henk Visscher.

Special thanks go to David Bond and Rob Newton, both excellent, multitalented research scientists and valued companions on many a field trip; we have spent many air miles together and a lot of time at Schiphol!

Last, I thank Jonathan Payne and a kind but anonymous reviewer of the original manuscript and especially Alison Kalett of Princeton University Press for sage advice and keeping me on topic during the many iterations of the pages that follow.

PROLOGUE

The public perception of mass extinctions is focused on the death of the dinosaurs, and for good reason. Stories do not get much more dramatic than the long reign of reptilian leviathans being abruptly terminated by a gigantic meteorite impact. However, this is not a book about that because, surprisingly, this was not the worst thing ever to happen to life. Bad though the dinosaur extinction was, something more devastating had happened nearly 200 million years earlier. The end of the Permian Period saw the loss of more than 90% of all species. Losses on this scale were soon recognized early in the study of geological history. More than 150 years ago, the English geologist John Phillips published a textbook, *Life on Earth: Its Origin and Succession*. It contained a chart documenting the ups and downs of the diversity of life that clearly showed major low points at the times that we now recognize as mass extinctions: at the end of the Cretaceous and the end of the Permian. But by the mid-nineteenth century, the idea that the history of life may record catastrophic events was not in vogue. Catastrophism had had its day; championed by Georges Cuvier in the earlier part of the century, it had been replaced by a uniformitarian view that changes were slow and gradual.

All this changed with the publication of a paper in 1980 that provided excellent evidence for an instantaneous cataclysm at the end of the Cretaceous. The scientists involved, led by the father-and-son team of Luis and Walter Alvarez, showed that sediments at the level of the mass extinction had high concentrations of iridium, a metal that is rare in the Earth's crust but more common in meteorites. Catastrophism was back on the agenda.

The 1980s saw a frenzy of activity on the Cretaceous-Tertiary mass extinction, which resulted in the discovery of the giant crater at Chicxulub, underlying the forests of the Yucatan Peninsula. Gradually geologists also began to turn their attention to other mass extinctions to see if they too were caused by impacts. Thus, the Permo-Triassic mass extinction finally began to receive some long-overdue attention. Eventually, tentative evidence for an impact at this time began to be found: a few grains of quartz that may or may not have been formed under high pressure at the impact site, rare traces of carbon molecules that allegedly could only be formed during impact, etc. None of this is convincing. In fact, the true culprit had long been staring geologists in the face, at least those geologists who had visited northern Siberia. The Siberian Traps record a scale of volcanism that is unparalleled in modern times, and by the early 1990s it became apparent that they had erupted at precisely the right time to be implicated in the Permo-Triassic mass extinction. As it happens, similar gigantic lava fields were also erupted during the Cretaceous-Tertiary extinction in the Deccan region of India. The addition of another volcanism-extinction link started to look like a pattern was emerging. And so it has proved.

If you read some literature you can easily get the impression that giant volcanic eruptions always cause global-scale catastrophes. Unfortunately, it is not as simple as that. Sometimes volcanism, even at its most voluminous, can be benign. At other times the eruption-to-kill strike rate is 100 %. This volume considers one such period, one lasting 80 million years, when the link between volcanism and extinction was perfectly one-to-one. It includes the Permo-Triassic mass extinction, which was just one of six such disasters, albeit the biggest, that coincided with a whole series of giant eruptions. The effect on life was profound. Each extinction-and-recovery phase saw the appearance of new species arising from the hardy survivors. By the end, there were very few species left that had been present at the start. Earth's newcomers must then have given a collective groan when the seventh giant eruption began—but this time nothing happened; life carried on. The causes of the mass extinctions and the reasons for the following non-catastrophes lie at the heart of this book. It has something to do with a supercontinent.

THE WORST OF TIMES

CHAPTER 1

A TIME OF DYING

If you could travel back in time 260 million years, you would find our planet had an unfamiliar geography. Nearly all of the landmasses were united into a single, giant continent. This was Pangea, and it stretched from pole to pole. On the other side of the world you would find a vast ocean, even larger than the present Pacific, called Panthalassa. Plunging into the ocean you would see some vaguely familiar groups—including mollusks, corals, and fishes—present in abundance, but as you strolled around the land, everything would look entirely strange. Large, lumbering, reptile-like creatures with faces covered in blunt horns ruled the world, and they crashed and blundered their way through vegetation composed of giant fernlike trees and conifers.

Despite the strange and superficially primitive appearance of terrestrial life, it actually represented a spectacular evolutionary achievement. This was the middle of the Permian Period, and for the first time, animals and plants had spread throughout the land and away from the wet habitats around rivers and swamps. This was the result of innovations, such as reptilian eggs and conifer seeds, that meant many organisms could now survive on dry land. In contrast, there had been few recent changes in the oceans. The Middle Permian marine realm was rather like that of the Carboniferous Period, and it was not a great deal different from that of the Devonian

Period before that. But this business-as-usual story was about to change. The first mass extinction in 100 million years was shortly to strike. The dominant land animals would be wiped out along with many of the ocean's most common species. This was a disaster, and it was just the first of a series of six catastrophes spread over the next 80 million years and included the worst examples the world has ever experienced. By the end of this age of extinctions, life everywhere had changed profoundly. In the oceans, the entire food chain, from the smallest plankton to the largest predator, was totally transformed. It was the same story on land. Dinosaurs now ruled the roost while swift little mammals darted around their feet. With the singular exception of the mass extinction that removed the dinosaurs, 66 million years ago, life was never again to experience such traumas.

This book attempts to explain and understand this worst 80 million years in Earth's history, a time marked by two mass extinctions and four lesser crises. To put the Pangean trauma into context, it is important to note that five major mass extinctions have afflicted the course of life. Scientists define mass extinctions as geologically brief intervals when numerous species go extinct in a broad range of habitats, from the ocean floor to forests, and at all latitudes, from the equator the pole. A true mass extinction represents global devastation with no hiding place. The first of the "big five" occurred 444 million years ago, at the end of the Ordovician Period. It was a fascinating event, associated with a short but intense glaciation (making it the only mass extinction event to be clearly linked to a cooling phase), and I wish I had reason to write more about it here, but it is not relevant to our story. Number two on the mass extinction list happened 70 million years later, during the Late Devonian Period. This was a time of several closely spaced crises for both marine life and also for the newly evolved amphibians, which had just taken their first steps onto land. Again, it predates the formation of Pangea and so is outside the scope of this book. Next up on the mass extinction roster is the Permo-Triassic crisis, and this is very definitely within the remit of this book. The gap between this crisis and the next, at the end of the Triassic, was the shortest of all the intervals between catastrophes, only 50 million years. Life did not have long to recover, and in fact, the Triassic Period was beset by its own succession of crises.

The final crisis was only 65 million years ago, and it famously wiped out the dinosaurs and many other groups, including the lovely ammonites, whose coiled shells make such attractive fossils. This Cretaceous-Tertiary crisis, as it is known, has been famously linked with a giant meteorite impact in the Yucatan Peninsula in Mexico and also with volcanism in India. Discussion of this event, and the debates on the cause, is also mostly beyond the remit of this book, although it gets a mention in chapter 7 because it provides a useful comparison with the older extinctions described here.

This book therefore includes two of the big-five mass extinctions in Earth history (the end-Permian and the Triassic) and four other extinction events. In so doing, the book attempts to put the subsequent success story in context and aims to provide an understanding of why life has since become so much more resilient, or at least much less prone to catastrophes (the occasional meteorite impact excepted) in the most recent 180 million years.

The time of interest begins in the middle of the Permian Period, spans the entire Triassic, and finishes in the Early Jurassic (fig. 1.1). There is no overarching name for this interval; on the contrary, it straddles one of the major divides of geological time, that between the Paleozoic and Mesozoic Eras, which are generally treated separately in geological and paleontological textbooks. This is unfortunate because the Permo-Jurassic has many recurring themes and similarities, and when viewed

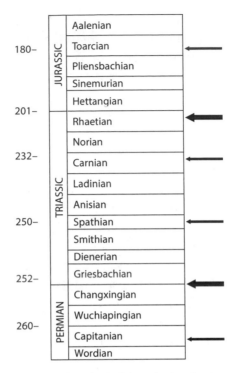

Figure 1.1. Subdivisions of geological time during the time of the Pangean supercontinent. The major extinctions are marked with arrows, and their age, in millions of years, is given to the left.

as a whole, it can be seen as a time when the Earth's oceans and climate showed distinctive and repetitive patterns. This is not to say that the interval is poorly studied. It includes the greatest disaster of all time, the Permo-Triassic mass extinction, 252 million years ago, which has been the subject of many academic papers and a substantial number of popular science books. The attention is merited because it was the world's worst ordeal. Its cause is one of the great topical debates in science. However, setting this mass extinction in its temporal context and comparing it with the lesser-known extinction events can help explain its origins and dispel any notion that it was a unique crisis. In fact, it was just the greatest of series of extinctions that had two factors in common: they occurred when the world's continents were united into the single continent of Pangea, and they coincided with gigantic volcanic eruptions. This book examines why volcanism at the time of a supercontinent is so bad for life.

Besides the great Permo-Triassic mass extinction, there were five other crises between 260 and 180 million years ago. Geologists only discovered some of these in the past few years, and so we are at an exciting stage with much to learn about them. The first extinction in the lifetime of Pangea, in the middle of the Permian Period, is generally known as the Capitanian extinction, being named after the geological interval in which it occurred. (Figure 1.1 lists the various time subdivisions used by geologists.) The crisis had a clear effect on tropical marine life, and it may have been devastating on land as well, as will be shown in chapter 2. There was only an 8-million-year recovery period from the Capitanian extinction before the devastating Permo-Triassic mass extinction struck. This caused entire ecosystems to disappear—it produced a world without forests and oceans without reefs. Vast swathes of the world were left devoid of life. The painfully slow recovery that followed in the first 5 million years of the Triassic was long regarded as a consequence of the sheer scale of the preceding blow. However, the latest research shows that there was probably another environmental calamity only a few million years after the Permo-Triassic mass extinction that knocked life back again before it had even begun to get back on its feet; this was the Smithian/Spathian crisis.

Only after the Early Triassic do we see a prolonged phase of diversification lasting more than 10 million years, the most peaceful interval of Pangea's history. This takes us to the Carnian Stage of the Triassic, when enigmatic and strange climatic change coincided with remarkable and equally puzzling changes among plants and animals. We are only just beginning to unravel the story of this time, let alone understand it.

Next up was the end-Triassic mass extinction, 202 million years ago. Once again huge changes were wrought on communities on both land and sea, with perhaps the most consequential being an emptying of the terrestrial landscape that allowed a formerly insignificant group called the dinosaurs to take center stage. This extinction marked the start of nearly 140 million years of almost trouble-free, dinosaur-dominated history. Or did it? Well, not quite. Within 20 million years of the start of the Jurassic, during the Toarcian Stage, a final extinction struck. It had many of the hallmarks of earlier Pangean crises, albeit with a much more muted expression. Its effects are best seen among marine life, whereas it is not clear if the dinosaurs (or anything else on land) were in any way bothered at this time.

After the Toarcian, Pangea began to split up and life thrived once again. The vicissitudes of times past were forgotten until one fateful day 66 million years ago, when a large meteorite struck the Yucatan Peninsula in Mexico. So, did life become tougher and less extinction-prone during the Jurassic, or did it just get luckier, with fewer environmental disasters? To answer this question we need to understand first how the Pangean extinctions were caused and then we need to know if these conditions were replicated latter.

The story of the worst 80 million years and its coincidence with Pangea also requires us to be familiar with the geological history of this supercontinent. Its assembly began before the Permian, when a large, southern hemisphere continent, called Gondwana, collided with a large, northern continent, called Laurasia, around 300 million years ago. The result was a larger continent (Pangea) with a mountain range, called the Central Pangean Mountains, along the suture that ran roughly

Figure 1.2. World map during the Permian Period, 260 million years ago, showing that most continents were united into the single supercontinent called Pangea. The other side of the world was single giant ocean called Panthalassa. Outlines of the modern continents are faintly shown.

east to west through the equatorial heartland (fig. 1.2). The worn-down segments of this range have now been split up, as a result of the formation of the Atlantic Ocean, but they can still be seen in the United States (the Allegheny Mountains), Morocco (the Anti-Atlas Mountains), and Spain, where they are impressive but not on the scale of the Himalayas.

Following the Gondwana/Laurasia collision, the final major pieces of the Pangean jigsaw—the continents of eastern Siberia and Kazakhstan—were the next to collide in the Early Permian and they formed the Ural Mountains. The final result of all this multicontinental pileup was a vast, arcuate supercontinent that by Middle Permian times, 260 million years ago, stretched from the North to the South Pole. The northern and southern arms of Pangea formed the shores of a large equatorial ocean called Tethys, while the other side of the world was entirely the truly vast Panthalassa Ocean. However, Pangea had not quite gobbled up all the world's continents: the eastern end of Tethys was partially blocked with several small continents that are today to be found in southeastern Asia. It is in the nature of small continents that they tend to be low-lying and so are commonly inundated by the sea. South China was one of these small continents, and thanks to persistent marine flooding, its marine sedimentary rocks provide an excellent record of life in the oceans.

No sooner had Pangea assembled than it started to fray at the edges as small continental slithers broke away from the Gondwanan margin and drifted northward through the Tethyan Ocean. By the end of the Triassic, 200 million years ago, these fragments (which included parts of present-day Iran, Turkey, and Tibet) had collided with northern Pangea. At the same time, the continents of North China and South China had similarly glued themselves to the northern Tethyan margin. The end result was a truly unified supercontinent that had its brief apogee in the Early Jurassic. However, continental drift is a ceaseless process, and fragmentation began immediately after the assembly. The first rupture began in the equatorial heart of Pangea, where Africa and the Americas were joined, and it spread southward as the Atlantic gradually "unzipped."

The mere existence of Pangea alone was not enough to create hostile conditions; indeed, for most of the continent's duration, life was constantly diversifying, and as we shall see, many new groups evolved, including the dinosaurs, mammals, and flowering plants. The key factor in the six crises of Pangea was volcanism. This was not the normal, everyday-type volcanic activity that produces volcanoes; rather, it was the most voluminous style of eruption ever recorded. Every Pangean extinction event coincided with the outpouring of enormous fields of lava called flood basalt provinces. The lavas were very low viscosity and flowed for hundreds of kilometers, infilling valleys and hollows in the land surface with a sea of magma. Successive flows stacked up in a series of thick layers that gradually weathered to form a staircase-like topography that is often called "traps," named after the Dutch word for "stairs" (and in English we have trapdoors that lead to staircases). Geologists also call these volcanic regions large igneous provinces, which allows them to use the acronym LIPs and thereby give semi-amusing titles to conference talks, such as "LIPs—The Kiss of Death" and "Beware of Big, Wet LIPs" and ... anyway, you get the idea.

It is probably fortunate that there is no volcanism today that approaches the scale of LIP volcanism. Each province typically includes at least a million cubic kilometers of lava composed of hundreds of individual flows, each with volumes of several hundred to several thousand cubic kilometers. No eruption in historical time has come anywhere close to being so large. For comparison, the Mount Pinatubo eruption of 1991, the biggest eruption of the twentieth century, involved only 5 cubic kilometers of magma, and even the Tambora eruption of 1815, probably the largest eruption of the past millennium, erupted only 30 cubic kilometers of magma. Clearly LIPs provide a very big "smoking gun" to explain Pangean mass extinctions, but explaining just how the volcanic "bullet" did the killing is far from understood. Making the connection between volcanism and catastrophe is made even more difficult by the fact that although the relationship

Pangea + LIP volcanism = mass extinction

holds true, once Pangea is removed from the equation, the link fails. By Cretaceous times (145–66 million years ago),

Pangea had long since broken apart, and the LIPs that erupted in this period did not cause major extinctions. Only the final LIP of the Cretaceous, the Deccan Traps of India, coincided with the famous death of the dinosaurs. But of course this crisis also coincided with a meteorite impact in Mexico, thereby vastly complicating all cause-and-effect scenarios. Further LIP eruptions have occurred within the past 65 million years, including a truly enormous example now found along the margins of the North Atlantic, but their consequences were fairly insignificant.

The task of this book is therefore to examine what happened during the Permo-Jurassic extinctions of Pangea, evaluate what may have caused these catastrophes (more specifically, to ask, how volcanism could have done it?), and finally to understand whether the resilience of the biosphere has changed in 260 million years or whether it has just become luckier thanks to continental separation; in other words, are supercontinents bad for life?

An incidental bonus of working on past environmental disasters is that giant volcanism produces effects that may be akin to modern anthropogenic activity, such as the emission of huge amounts of carbon dioxide into the atmosphere. The relevance of understanding ancient "violent shocks" when it comes to predicting near-future worlds has not been lost on geologists, not least for the prosaic reason that it provides a justification for research funding.

Until recently, most modern extinctions could be attributed to overhunting or habitat destruction. One need only think of the iconic dodo or the numerous bird species that were lost from Hawaii as the forests were destroyed and exotic animals introduced. However, the past decade has witnessed rapidly growing concerns over the effects of global warming and climate change. The rapid shift of climate belts may prove too fast for species to migrate, especially in fragmented landscapes, where only small islands of original habitat survive in a "sea" of agricultural land. In the oceans, warming will cause acidification by carbon dioxide and a decline in dissolved oxygen. Both factors are major contenders, albeit contentious ones, in marine extinction hypotheses for the Pangean crises. The debate on what caused the Pangean extinctions is therefore of relevance to the attempts to predict how life will survive in the coming centuries, although I do not intend to overstress this latter aspect. This is primarily a book about a time when Earth was very different, a time of supercontinents, super-oceans, and super-eruptions, and above all, an age of mass extinctions.

CHAPTER 2

EXTINCTION IN THE SHADOWS
THE CAPITANIAN CRISIS

The world map of 260 million years ago looks very strange (see fig. 1.2).The presence of a single continent meant it would have been possible to walk from the North Pole to the South Pole without ever getting your feet wet. Passing through equatorial latitudes would have required a climb over a Himalayan-scale mountain range, but most of the time it would have been a slog through desert plains. You might have encountered glaciers in the highest mountains, but there were no ice caps at either pole. Instead, the polar latitudes were the sites of remarkably luxuriant forests, especially in the southern hemisphere, where the thick coals that are now of enormous economic value were forming in South Africa, India, and Australia. Lush equatorial swamps could also be found, especially in China, where the lycopsids thrived. These were the first tree-sized plants to evolve, and they are still with us today, but only as small plants called club mosses, which still prefer to grow in wet, swampy conditions. In other regions the common trees included the pteridosperms (seed ferns), of which the glossopterids (in southern polar regions) and the gigantopterids (in equatorial latitudes) were the most abundant. Another major plant group, the gymnosperms, with their naked seeds and large woody trunks, were the latest word in evolution in the Permian. Their heyday lay tens of millions of years in the future, but even in the Permian, the cordaite gymnosperms were important players in northern latitude forests.

Walking among the trees was a diverse range of animals that, for the most part, belonged to a group called the dinocephalians. This name translates as "terrible heads," in recognition of the bizarre lumpy and horned skulls of these animals. They comprised impressively large animals and included large predators up to 6 meters in length and herbivores comparable in size with modern hippos. They were by the far the largest animals to have walked on land up to that time. Alongside the dinocephalians there was a menagerie of insects and arthropods that included familiar types like scorpions and cockroaches, true methuselahs of the fossil record that are still with us today, and less familiar groups, including huge dragonflies and giant herbivorous forms called the palaeodictyopterids, which had specialized mouthparts formed into a sucking beak.

Life on the seabed encompassed a similar mixture of the strange and the familiar. A few modern groups, such as bivalves (clams) and gastropods (snails), were present, but the most abundant types, such as the brachiopods and crinoids, would eventually become either rare or totally extinct in modern oceans (e.g., trilobites and several coral groups). Brachiopods look a little like clams because they have two valves held tightly together, but the resemblance is superficial. Today they can still be found in a diversity of places, such as the shallow waters around New Zealand, but you will be lucky to find their shells on a beach. In contrast, in the Permian they were among the most abundant shelly organisms, as were the crinoids. This group of animals, having a many-branched head on top of a long thin stalk, is also still around, and their common name is sea lily. They are mostly found in deep waters, although some species live on coral reefs. In the Permian the crinoids formed dense forests wafting in the currents. Swimming above these bottom dwellers were nautiloids and diverse ammonoids, distant relatives of the modern squid, with coiled shells. Alongside swam a diversity of fish, including sharks, not greatly dissimilar to those of today, whereas others were actinopterygians and holocephalians. The former resembled modern fish but were not particularly closely related, and the latter were odd-looking durophagous (shell-crushing) types.

Until recently, there were two prevailing viewpoints concerning the final 10 million years of the Permian life: either it had continued to thrive until an end-Permian fin de siècle, or it had declined gradually in response to deteriorating conditions. Two parallel studies abruptly changed this perception. Jin Yugan of the Nanjing Institute of Geology and Palaeontology (NIGP) was the first to announce discovery of a Middle Permian extinction in a 1993 talk that was published in 1994. Nineteen ninety-four also saw the appearance of an independent paper by Steve Stanley (then at John Hopkins University) and Yang Xiangning (then of Nanjing University) that reached much the same conclusions.

Both teams had compiled lists of the known fossil occurrences in the Middle and Late Permian and revealed two extinction peaks: the well-known one at the end of the Permian and an earlier peak, at the end of the Middle Permian (which is also called the Guadalupian), both separated by an interval with lower amounts of extinction (fig. 2.1). Stanley and Yang's study particularly highlighted the losses among fusulinaceans. These were a group of foraminifers (usually referred to by their abbreviated name of foram, as they will be here), which are single-celled protists that secrete tiny, chambered shells. Forams have an abundant fossil record throughout the past 400 million years, and they can still be found in marine

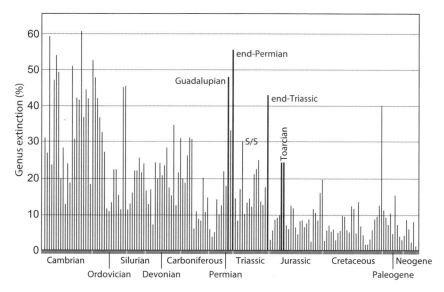

Figure 2.1. Extinction rates since the start of the Cambrian Period, 540 million years ago, to the present day.

sediment, although you need a microscope to see them. The fusulinaceans were an unusually large and immensely diverse subgroup of forams that could reach sizes of up to 10 centimeters in length, although they were typically a centimeter in size (fig. 2.2). They preferred warm waters and so were very common in shallow, tropical seas. More than three-quarters of the fusulinacean genera disappeared, and they are the signature extinction group for an event that has come to be known as the Capitanian mass extinction—the Capitanian being the final interval of the Guadalupian (fig. 1.1).

The fusulinaceans were not the only losses of the Capitanian; many other species in the tropical seas also vanished, including corals, brachiopods, crinoids, and ammonoids. The corals are familiar today, but the Permian examples belonged to distantly related forms groups known as tabulate and rugose corals, which constructed their skeletons in a distinct

Figure 2.2. Fusulinid foram: a giant single-celled organism around 2 millimeters across. They thrived in warm shallow seas until their virtual extermination during the Capitanian mass extinction.

and different way. Brachiopods have a good fossil record because they secrete two robust valves of calcite. Their fossils provide some of the best evidence for the Capitanian extinction story, especially in the shallow-water seas covering South China, where 90% of brachiopod species failed to survive the Capitanian. The ammonoids also suffered similarly grievous losses, although for this group it was not so unusual for them to suffer a catastrophe. Ammonoids finally disappeared at the end of the Cretaceous, but this was only the last of a series of extinctions that punctuated their nearly 200-million-year existence. Their entire history was a roller-coaster ride of cat-astrophic extinctions and spectacular recoveries; this Mid-dle Permian crisis was just one of the many calamities they suffered.

Within a few years of its discovery, it was clear that the Capitanian marked a crisis that justified Stanley and Yang's conclusion that it "would long ago have been recognized as one of the great mass extinctions . . . if its effects had not been lumped with those of the terminal Permian event." Remark-ably, within a year of its discovery, a potential culprit was dis-covered, a contemporaneous giant volcanic province called the Emeishan Traps.

Like the discovery of the Capitanian extinction, it is rather surprising that something as big and obvious as an immense region of basalt lava should only have come to light as late as the 1990s. Where had it been hiding? The answer is mostly in southwestern China, with a little bit in northern Vietnam. The reason for its tardy discovery lies not only in its remote location (for western geologists), but also because, since the Permian, it has become badly mangled by tectonic activity. The collision of India with Asia around 45 million years ago caused intense mountain building, with the result that the Emeishan lavas became uplifted, folded, faulted, and then eroded; the main fragments that remain are now found in mountains in the Chinese provinces of Yunnan, Sichuan, and Guizhou. No other LIP of Pangea has been so broken up and eroded. As the ages of the Emeishan fragments were gradually determined, it became apparent that they were once part of a continuous region of lava of considerable size. Initial dating of the lavas suggested that they coincided with the Permo-Triassic boundary and thus could have caused the Permo-Triassic mass extinction. This link between volcanism and extinction was the right idea, but it was wrong extinction. In fact, even the earliest evidence showed that the Emeishan

LIP was older than the Permo-Triassic boundary because the lava pile is overlain by sedimentary rocks that contain Late Permian fossils. Vincent Courtillot, a major player in French academia and unarguably the most ardent proponent of LIP/ mass extinction links, was the first to correctly note in his fascinating book *Evolutionary Catastrophes* (originally published in French in 1995) that this newly discovered large igneous province coincided with the new Middle Permian mass extinction.

The next few years saw increasing attention given to the Emeishan LIP, revealing it to be rather a small example of its kind: the original volume has been estimated at less than a million cubic kilometers, making it a lot smaller than some of the later provinces of Pangea. Thus, within a few years of its discovery, some geologists were beginning to have second thoughts about linking the Capitanian extinction to the Emeishan volcanism. Yukio Isozaki of the University of Tokyo was particularly unimpressed by the province's size and suggested in 2001 that some other volcanic event at an unspecified location caused the end-Capitanian extinction. It was at this time that I decided to get involved in the study of this crisis. I had just finished writing a review of mass extinctions and LIPs (inspired by Courtillot's book), which had concluded that, among other things, trying to make the connections between these two phenomena was bedeviled by imprecision in timing.

Geologists are generally happy if they can show that past events happened within a few million years of each other, but for mass-extinction studies, such resolution is not good enough. It is particularly difficult to achieve dating precision if events happened in different parts of the world. Fortunately, the Emeishan volcanism has one superb advantage when it comes to trying to unravel what happened: the volcanic and fossil records are found in the same place and therefore can be compared directly. The lavas were erupted into a shallow tropical sea in southwestern China where carbonate muds (now limestone and called the Maokou Formation) were accumulating. The sea was full of the usual Permian life, with the result that the Maokou Limestone is richly fossiliferous. Even better, the limestones did not stop accumulating with the onset of volcanism, but rather they continued to form during the intervals between eruptions. Thus, the interbedded layers of lavas and limestones could potentially provide both the history of volcanism and extinction in the same locations—and so it proved. All that I needed was to assemble an "A team" of geologists and paleontologists to help me study the rocks.

First on the list was Jason Ali, who had already visited the Emeishan Province, where he had been studying its paleomagnetism. Jason is a paleomagnetist, although other geologists usually call them "paleomagicians," owing to the mysteries of their dark art. He is an enthusiastic proponent of this art, which involves measuring the alignment of the magnetic minerals in rocks such as lavas. Magnetic minerals align parallel with the Earth's magnetic field and so have a north-south orientation, but the field regularly flips, and magnetic north becomes magnetic south. The frequency of these reversals, as they are known, is typically every few million years, but the interval between 312 and 265 million years ago is highly unusual because the magnetic field remained stubbornly stable at this time; there were no reversals at all. This interlude has been called the Kiaman Reverse Superchron ("reverse" because a compass needle would have pointed toward the south pole at the time), and it is important to our story because the Emeishan eruptions began not long (in geological terms) after the superchron ended, thus providing a useful datum. It is especially important when trying to correlate terrestrial and marine sedimentary rocks because they can both preserve a record of reversals.

A more traditional way of dating sedimentary rocks involves using the fossils they contain: the fossil of choice for Permian rocks is the conodont, a tiny, spiky, little piece of bone that used to be in the jaws of the conodont animal. The latter are now extinct but are thought to have been primitive eellike fish barely an inch in length. Conodonts fossilized easily, and fortunately for paleontologists trying to age-date rocks, they come in many different types because they evolved and changed rapidly. Thus, it is possible to date rocks to very high resolution using these fossils. Conodont paleontologists are the equivalent of archeologists who use coins to tell the age of Roman remains. If the Romans had used only one type of coin throughout the duration of their empire, then they would be useless objects for dating, but because the coins changed rapidly, with new issues every time a new emperor came to power, they provide a detailed chronology. Conodonts provide a similar time series, but extracting them from rocks is not straightforward; you first need to collect a lot of rock, which then has to be dissolved with strong, unpleasant acids, and then the conodonts have to be separated from all the residue. This part is done by picking them out with the moist tip of a fine paintbrush while looking down a microscope—a time-consuming job. Fortunately I knew just the man to do this—Lai Xulong, of the China University of Geosciences in Wuhan. Xulong and I had already collaborated on several mass extinction projects, and I am always happy when he is onboard with a project. His vast network of personal connections in China ensures that fieldwork always runs smoothly, with regular banquets along the way. He is also blessed with the ability to recruit top-notch research students, and for this project we had Jiang Haishui and Sun Yadong, both of whom would soon be busy extracting conodonts from the Maokou Limestone.

Jason and Xulong's team provided the expertise for dating the Emeishan lavas and limestones, but other attributes also needed to be evaluated if we were to get a full picture of the environments of the Middle Permian. Thus, Jason Hilton of the University of Birmingham (in the UK) provided expertise on the fossil plant story and ensured that we had a good quota of Jasons on the project. Dave Bond and Stéphanie Védrine, two young postdoctoral researchers working with me in Leeds, covered the sedimentology and fossil angles, while Wang Wei of the Nanjing Institute of Geology and Paleontology and Rob Newton, another colleague at Leeds University, provided expertise in geochemical analyses. With the team assembled, we descended on southwestern China, hammers at the ready. What was planned was a thorough study of the Emeishan eruptions and the associated environmental changes. Extending our studies beyond China, we were also interested in seeing how the crisis played out in the rest of the world.

Our initial fieldwork began beyond the northern periphery of the lava fields in Sichuan Province (famous for its spicy food) and looked at the limestones of the Middle and Upper Permian, where the only manifestation of the eruptions was a thick layer of volcanic ash, called the Wangpo Bed. Just below this level there is a break of some unknown duration in the limestone accumulation—a first job for Xulong's conodont-extraction team. Their results revealed a large time gap (known as a hiatus) spanning much of the Capitanian Stage. This is unfortunate but at the same time potentially important. Unfortunate because it means that there is a long period of time when we have no record of what was happening in the local environments of Sichuan, and significant because of what may have caused the hiatus.

Limestones typically accumulate in shallow, warm seas (the Maokou limestones of Sichuan are no exception), but they stop forming if the environment becomes emergent, which can happen if the sea level falls or the seafloor is uplifted. So what happened in Sichuan? The presence of the

Emeishan LIP to the south may provide the answer. Large igneous provinces come from mantle plumes, which are large columns of warm, buoyant mantle that arise from the core-mantle boundary, deep within the Earth, and slowly ascend to the surface. When they reach the base of the crust, they are initially thought to cause doming and uplift before they break through and pour out great volumes of magma that become the LIP lava flows. Thus, the break in deposition of the Maokou Limestone could record the arrival of a mantle plume beneath the South China continent that caused crustal doming before eruption. The Emeishan Province has become something of a type example for this important geological phenomenon. However, for our purposes we needed more continuous records of the interval around the time of onset of Emeishan volcanism, and so we headed south from Sichuan to Guizhou and Yunnan Provinces.

In a mountain valley near Xiong Jia Chang, in the remote western border region of Guizhou Province, there is a dusty roadcut and alongside is a fast-flowing stream running over bedrock. After heavy rainfall the stream level is too high to examine the rocks, but when it is drier, the water-worn rocks reveal a remarkable transition from a tranquil seafloor to a time of unimaginable explosive violence. Inauspicious as they seem, the rocks of Xiong Jia Chang provide one of the most detailed records of what happens when a giant volcanic province begins to form in shallow tropical seas. Jason Ali led us to the location; he had recently obtained an excellent magnetostratigraphic record from there, and the rocks were to yield further information. The order of rocks from bottom to top at Xiong Jia Chang is as follows: limestone—lava flow—limestone—lots of lava flows. Thus, the start of eruptions briefly shut down limestone production, only for it to start up again afterward before the shallow carbonate seas were then finally swamped by a whole succession of flows.

Jason's magnetostratigraphy showed that all these events occurred shortly after the Kiaman Reverse Superchron, a good clue to their Middle Permian age, and our age control became even better once Xulong, Yadong, and Haishui retrieved conodonts from the limestones. Thus, we were able to show that first blast of volcanism occurred within the time zone defined by a conodont called *Jinogondolella altudaensis*. Limestone deposition resumed after the first volcanism in the next younger conodont zone that contained two more species of *Jinogondolella*, which give their names to this time interval: the *J. prexuanhanensis—J. xuanhanensis* Zone.[1] After the formation of the *J. xuanhanensis* limestones, the Emeishan volcanism started again on a big scale, and more than 200 meters of pyroclastic rocks and lava flows piled up at Xiong Jia Chang. To the west of this location, in Yunnan Province, this volcanic pile reaches a peak thickness of nearly 5 kilometers.

The Xiong Jia Chang section provided crucial data on the timing of volcanism that then allowed us to see what happened to life in the Maokou seas, both at this location and elsewhere in South China where the same conodonts are found. The fossils in the limestones below the lowest volcanic flow are typical of Capitanian assemblages; they include fusulinacean forams, along with brachiopods, sponges, and calcareous algae. When limestone deposition resumed in the *J. xuanhanensis* Zone, after the first phase of volcanism, the fusulinaceans had gone, along with many other types of fossil. This could have reflected merely a local change, but comparison with fossil records elsewhere in South China showed that the losses were genuine disappearances. Just as indicative that something bad had happened was the composition of the post-eruption creatures: they had low diversity and were dominated by a foram called *Earlandia*, which was an ultrasimple form, just a narrow cone. The dominance of simple forams present in large-abundance-but-low-diversity fossil assemblages is a common phenomenon in the aftermath of most mass extinctions; we will encounter such assemblages again. Thus, *Earlandia* also bloomed in the immediate aftermath of the end-Permian mass extinction. The ability to thrive at such times has led paleontologists to call fossils such as *Earlandia* "disaster taxa" (note that the term "taxa" is a biological word for any groups of organisms of species rank or higher).

With extinction victims and post-extinction disaster taxa both located, we could be confident that we had precisely found the mass extinction interval, and crucially, it was immediately below the first flow of the Emeishan LIP. Equally important, the extinction level was not at the end of the Capitanian as previously thought, but several million years older.

So, why had earlier studies claimed a younger date for the extinction? The answer relates to the way the mass extinction was discovered in the first place—on the basis of data compiled from the paleontological literature on Middle Permian fossils. Most such publications do not give the precise age of fossils but simply attribute them to a stage. Stages are the subdivisions of geological time below the better-known periods, and they can have highly variable durations, anything from 1 to 20 million years. Anyone attempting to understand how long a fossil lived generally makes the assumption that if a species is found within a stage but not in the next one, then it must have gone extinct at the end of the first stage. This supposition is sensible, but it causes problems because it does not help determine when in the stage the extinction occurred. This approach is analogous to a historian in the distant future piecing together the military casualties of the twentieth century and discovering that there were tens of millions of deaths. In the absence of any other evidence, he or she then makes the assumption that all the deaths in the twentieth century happened at the end of 1999 during some cataclysmic end-of-century battle. Obviously this is a foolish thing to do, but in the absence of any other evidence, it is all that can be done. Paleontologists compiling extinction inventories from published records are faced with this problem, and the concept of an end-Capitanian mass extinction derives from a similar situation. Such conjecture is routine in most paleontological studies, and in some cases it is probably justified. Many extinction events do indeed appear to have occurred at the end of a stage, but it need not always be so. The only way to avoid the everything-died-at-the-end practice is to do fieldwork, collect fossils, and compile one's own fossil record, as we did in southwestern China.

Having precisely identified the interval of marine extinctions, it was then important to ascertain if anything comparable had happened to terrestrial life. Terrestrial animals showed rapid evolution throughout the Permian as several short-lived dynasties came and went. Plants were also diversifying rapidly and, to modern eyes, had a rather mixed old-and-new composition. With the benefit of hindsight, the Permian forests can be seen to have been undergoing a transition from lycopsid-dominated forests, which were characteristic of the preceding Carboniferous, to the gymnosperm-dominated forests of the Triassic, and for many years paleobotanists considered this changeover to be a very gradual one. However, the Permo-Triassic mass extinction has more recently been shown to record a major hiccup in plant evolution (as related in chapter 3), and the latest research suggests that the Capitanian may not have been a happy time for plants either.

The best Middle Permian fossil plant record is in North China, where Jason Hilton and his student Liadan Stevens have been studying the sandstones and shales of the Shihhotse Formation. These rocks were deposited in a mixture of river and lake environments, optimum conditions for preserving fossil plants. It is therefore surprising that the entire flora of more than 40 species rapidly disappears in the upper part of this formation. There appear to be no local environmental changes recorded by the sediments, so could these losses instead have been a response to some extraregional stress—Emeishan volcanism perhaps? A first step in testing this idea would be showing the coincidence between eruptions and the loss of the plants. Dating the Shihhotse Formation is achieved using the magnetostratigraphic technique described above. This procedure revealed that it occurred around 3 million years after the end of the Kiaman Reverse Superchron and thus within the middle of the Capitanian Stage. Perfect! The plant crisis and the marine extinctions therefore both coincide with Emeishan volcanism. The Stevens and Hilton results are encouraging and suggest that the Capitanian crisis was a true mass extinction affecting both marine and terrestrial communities, but it is unsafe to infer a global crisis based only on evidence from China. Let us look elsewhere.

Capitanian rocks formed in terrestrial environments are also found in southeastern Spain, and they too record a major plant crisis. The result was the loss of densely vegetated peat mires, which meant that coals had ceased to form. This loss is also associated with an intriguing change in the style of rivers in the region. Prior to the plant crisis, the rivers were slow-flowing, meandering channels, but they were replaced by swift-flowing, braided rivers characterized by lots of parallel channels separated by small temporary islands. As we shall see, rivers also showed this meandering-to-braided transition in the immediate aftermath of the Permo-Triassic mass extinction. This behavior provides a potential signal of plant dieback because meandering rivers are characteristic of heavily vegetated areas where plant roots provide bank stability and so cause rivers to flow in lazy meanders through relatively narrow channels with stable banks. In contrast, rivers flowing through weakly vegetated landscapes are much straighter, more numerous, and more prone to switching their course frequently.

So, if the mid-Capitanian crisis affected terrestrial vegetation, then it is not unreasonable to expect that terrestrial animals also suffered. Not surprisingly, there was indeed a devastating extinction: all the dinocephalians, large and small, herbivore and carnivore, abruptly disappeared to be replaced by several rapidly evolving new groups, especially the dicynodonts and gorgonopsids. The former would go on to produce diverse herbivores in the Late Permian, and the gorgonopsids were carnivores that ate them. The dicynodonts were stout-bodied herbivores with blunt faces and a distinctive mouth that had tusks and a snipping beak, which was no doubt used for slicing off vegetation. The gorgonopsids, on the other hand, looked sleeker and had jaws full of sharp teeth. All these Permian groups—the dinocephalians, dicynodonts, and gorgonopsids—belonged to a larger grouping called the therapsids, which would, much later on (in the Triassic), give rise to a successful group called the mammals.

Reptiles were also present in the Permian, and in the later part of the period, after the demise of the dinocephalians, a large herbivorous group called the pareiasaurs became successful. They had large, fat bodies, a relatively small head, and leaf-shaped teeth similar to those found in modern herbivorous lizards such as iguanas. It is likely that they grazed vegetation alongside the dicynodonts although, strangely, their front legs were a lot longer than their back ones, thereby raising their heads some way above the ground. Presumably they ate leaves from small trees rather than ground-level plants.

Evidence for the dinocephalian extinctions comes from two widely separated parts of the world, the Karoo Basin of South Africa and the Ural Mountains in Russia. Surprisingly, the available age-dating for rocks in these basins suggests that the dinocephalian mass extinction happened several million years before the mid-Capitanian crisis. Indeed, it does not even seem to have happened during the Capitanian at all but in the preceding Wordian Stage. This is more than a little strange. Why are the plant and animal extinctions so out of step? It suggests that the animals died first and then recovered and that the plants went extinct several million years later, during their own separate crisis, at the same time that the animals were recovering. Given that the fortunes of the animals were closely tied to the plants, which they ate, one would expect a much more closely linked record. A simpler story would be to have synchronous plants and animal extinctions, but this explanation requires substantial reevaluation of the dating of the animal extinctions. The age of the dinocephalian extinction is based on magnetostratigraphy, and the evidence (mostly from Russia) seems to suggest that these animals disappeared before the end of the Kiaman Reverse Superchron. However, I have to admit to being skeptical about this mismatch between extinction timings. Collecting vertebrate fossils and undertaking magnetostratigraphic studies is rarely done simultaneously, and trying to compare two very different types of data collected at different times and locations can lead to dating error.

Greg Retallack of the University of Oregon has also been keen to realign the timings of the plant and animal extinctions and to ascribe a Capitanian age to the terrestrial crisis. Together with his colleagues, he studied the Karoo Basin's fossil pollen and spores alongside the vertebrate fossils and showed coincident extinction losses in both. In other words, both the plants and animals seem to have suffered a crisis at the same time. In addition, he also reported the same change in river style seen in Spain, from meandering to a low-sinuosity, braided style. This was all well and good; finding that both animals and plants suffered a synchronous crisis clearly makes sense. However, Retallack further suggested that the terrestrial crisis coincided with the marine extinction that, at the time of his study in 2006, was dated to the end of the Capitanian Stage. This age assignment of the terrestrial record was little more than a guess in 2006, and it required the Karoo rocks to be more than 7 million years younger than previously thought. Not surprisingly, Retallack's conclusions were severely criticized by other geologists. However, subsequent redating of the marine extinction to the middle of the Capitanian has narrowed this age discrepancy substantially, and I suspect Retallack's hunch that marine and terrestrial animals and plants died at the same time is correct.

So, by 2009, fifteen years after it was first discovered, the Capitanian extinction is now considered a true global crisis deserving of the epithet mass extinction. That it also coincided with the onset of eruptions of a giant-scale volcanism points to a potential culprit. However, the mantra "coincidence is not causation" is never far from the minds of those studying mass extinctions. Further work on the nature of the Emeishan eruptions and their potential environmental consequences, combined with what we can glean from studies of modern volcanism, has helped strengthen this cause-and-effect relationship.

The Emeishan volcanic rocks hold some potential clues as to their lethality. When assembling the original team, I had assumed (naively as it turned out) that the volcanic rocks would be easy to study. They were reportedly basalts, which are typically dark, fine-grained massive rocks—pretty straightforward rocks to study in the field. However, in China such lava flows proved to be the exception; the majority of the rocks were breccias—an ill-assorted jumble of angular rock fragments consisting mostly of lava but also including white blocks of limestone. The range of sizes of the fragments was enormous, everything from a few millimeters to many tens of meters. One large quarry we visited in Yunnan Province seemed to be purely limestone, but our scrabbling around its periphery showed us that we were looking at a single, giant block embedded in volcanic breccias. The presence of such gigantic boulders raised the question, how big were the beds they were found in? The answer proved difficult to assess because the exposures were rarely large enough to see both the base and top of the beds. At one location, a roadcut in northern Yunnan near Lugu Lake, the volcanic strata had been tilted up by tectonic movements. As a result, it was possible to walk along the road and gradually pass through higher and higher levels in a flow. I began doing this and had gone some way before I realized it would be quicker to go back, get in our field vehicle, and drive along the entire roadcut. However, by the end of the cut I had still not reached the top of the breccias, and so I could only estimate that the flow was much more than 100 meters thick and could easily be twice that thickness. The breccias must therefore record a phenomenal style of eruption.

After a few field seasons in the Emeishan, it soon became clear that I needed to call in expert help—bona fide volcanologists were needed—and so, with the promise of fieldwork in far-away places, Mike Widdowson, then of the Open University (located in central England), and Dougal Jerram, then of the University of Durham, joined the team. Both Mike and Dougal have vast experience in studying flood basalt provinces; Mike's career has concentrated on the Deccan Traps in India, and Dougal's travels have taken him to many LIPs. They soon identified that we had a style of eruption called phreatomagmatic on our hands in China. Such eruptions happen when volcanism occurs in shallow seas. A magma chamber just beneath the seabed runs the risk of seawater penetrating along cracks and reaching the hot molten rock, where it instantly turns to steam and expands explosively; the result is—kaboom! The violence of the Emeishan eruptions blew the Maokou Limestone seabed to smithereens, allowing further water to pour into the magma chamber to produce ultraviolent explosions of rapidly expanding steam and fragments of magma. One of the best examples in recent times, albeit on a small scale, occurred in 1963 south of Iceland, where a series of explosive eruptions produced a new volcanic island christened Surtsey.

The other common volcanic rocks to be seen in the Emeishan Traps are pillow lavas. Contrastingly, these result from one of the most peaceful styles of eruption. Pillows form as magma quietly oozes out of seabed fissures and rapidly cools to form a series of rounded blobs. Type "pillow lavas" into YouTube and you will see some nice examples forming in shallow Hawaiian waters and filmed rather closely, but ultimately safely, by an intrepid diver. In contrast, filming a phreatomagmatic eruption close up would require the diver to post his film from heaven. The contrast between these two shallow-water eruptive styles simply depends on the strength of the roof of the magma chamber. If it is readily penetrated by water, the result is spectacular, but if not, then the magma just dribbles out.[2]

The violence of the Emeishan volcanism may provide a clue to the cause of the mass extinction. Large and violent eruptions propel ash-laden clouds and gases into the atmosphere, where, like all clouds, they have a cooling effect. Chlorine and fluorine, two especially noxious halogen gases, are also emitted in quantity during volcanic eruptions, but volumetrically, sulfur dioxide (SO_2) is more significant. Basaltic magmas are rich in this acidic gas, and the eruption of 1000 cubic kilometers of basalts (probably a reasonable estimate for an Emeishan flow) would release around 10 gigatonnes. Note that a gigatonne can also be expressed as 10^9 tonnes, or 10^{15} grams, or a million billion tonnes. Anything measured in gigatonnes is clearly a large amount, and 10 gigatonnes is far greater than the annual amount of SO_2 currently reaching the atmosphere from fossil-fuel burning.

In the atmosphere, sulfur dioxide rapidly combines with water vapor to form clouds of sulfate aerosols, which block out sunlight. Large modern eruptions, such as that of Mount Pinatubo in June 1991, have a noticeable cooling effect that lasts for around a year or so, until rainfall removes the ash and aerosols. Because of the gases' short time in the atmosphere, the cooling effect is felt beyond the volcanic region only if they are injected into the stratosphere, where stratospheric winds rapidly circulate them around the globe. This requires the eruption column to reach heights of 5 to 6 miles above the volcano; a height only achieved by truly violent eruptions. Lesser eruptions spill their contents into the troposphere, where intense rainfall rapidly removes the ash and aerosols in a matter of months. Even aerosols in the stratosphere are only present for a few years, which barely gives them time to circulate from one hemisphere to another. Consequently, because of the aerosols' short time in the atmosphere, most violent volcanic eruptions are capable of causing only short-lived cooling episodes in one hemisphere. So the cooling effects of volcanic gases are generally thought to be very brief; however, they can have another unpleasant effect. The rainout of sulfate and halogen aerosols is acidic, and this is another cause of damage from volcanism, but again this acid-rain effect is unlikely to have global reach unless the gases can reach the stratosphere and circulate around the world.

So, could the gigantic, phreatomagmatic-style eruptions seen in the Emeishan Province have triggered a short, sharp cooling event that was sufficiently intense to cause extinction? The timing of the extinctions—at the start of the eruption sequence—is certainly suggestive that the first volcanic blast did the damage. Unfortunately (or should that be "fortunately"?), no modern eruptions have approached the scale of Emeishan volcanism, so we have nothing that directly compares to it. However, by looking a little further back in time we can find some impressively large eruptions. One such was the eruption of Toba, in present-day Indonesia, around 74,000 years ago. The violence of this event was sufficient to blow the volcano to pieces and scatter nearly a thousand cubic kilometers of ash over a large area. Despite its magnitude, it is unclear if Toba's eruption had any long-term affect on climate. Part of the problem is that the past few hundred thousand years have seen huge climatic oscillations associated with the waxing and waning of ice ages and all sorts of other short-term rapid climatic oscillations. Among all this climatic noise, trying to attribute any climate change to Toba volcanism is tricky. Nonetheless, the effect of this eruption may have been severe. Based on evidence from our own genetic code, it has been suggested that there was a severe contraction of global human population size around 70,000 years ago, tantalizingly close in age to the Toba eruption. Chimpanzee DNA similarly suggests a contraction in their population sizes about the same time. True extinction at this time has, however, not been recorded, leaving the value of the Toba eruption as an analogue for past LIP-induced cooling-extinction links highly questionable.

Because we have nothing comparable to Emeishan eruptions, estimates of volumes of released gases are based on extrapolations from modern eruptions and on the assumption that they are somehow comparable. There may be something we are missing about the nature of LIP eruptions—an "unknown unknown," as Donald Rumsfeld would say. Henrik Svensen of the University of Oslo has proposed a possible missing factor. Around a decade ago he was studying the North Atlantic Igneous Province, a giant region of basaltic rocks that formed approximately 60 million years ago. Images obtained by seismic surveying off the shore of Norway showed that hundreds of small craters had formed on the seabed directly above the magma intrusions that had fed the volcanism. Today these craters are buried beneath layers of younger sediment, but seismic analysis allows the volcanic structures within rocks to be easily discerned. Svensen interpreted the craters to be the product of explosive gas escape and postulated that the gas did not come from the magma itself but rather from the sedimentary rocks into which the magmas intruded. His argument is as follows. Basaltic magma is hot, having temperatures that are well above 1000°C, and when it comes into contact with rocks, it starts to bake them, in a process geologists call contact metamorphism, to produce a zone of cooked rock. Now if the rock contains organic matter, then the baking converts it to methane. This methane is said to have a thermogenic origin (it is produced by cooking), and it provides an additional source of gases associated with volcanism. The Norwegian magma intrusions did indeed come into contact with some very organic-rich rocks as they ascended to the surface, and so, Svensen postulated, there was the release of a huge amount of thermogenic methane to the atmosphere via hundreds of gas-escape pipes.

Svensen's hypothesis is important because it provides a way for LIP eruptions to inject extra gas into the atmosphere, not just that which bubbles out of the lava, thereby greatly escalating the environmental consequences. In the case of the Emeishan lavas, they passed through mostly organic-poor limestones and dolomites, and so we would not have expected them to generate any methane. However, baking such carbonate rocks would have released a great deal of carbon dioxide, leaving behind a cooked rock called marble.

The "plumbing system" beneath the Emeishan lavas consists of a network of horizontal and vertical intrusions called sills and dykes, respectively. Some of them are very big. The Panzhihua Sill is a particularly impressive example. Around 2 kilometers thick, it baked a zone of limestone into marble more than 300 meters thick. The amount of carbon dioxide released by this baking would have been impressive, a figure of 22 gigatonnes of carbon dioxide (CO_2) was probably generated from the baking effects of this sill alone. There are many others.

Our understanding of the cooling caused by the release of sulfur dioxide and halogens during massive volcanism suggests that the effect was very brief, so perhaps we should turn our attention to the long-term warming effects of carbon dioxide emissions. How significant would the warming be? Based on scaling up the gas release measured during modern eruptions, we can calculate that a 1000-cubic-kilometer Emeishan basalt flow (a rough estimate for the size of an individual flow) would have released about 13 gigatonnes of CO_2. Combining this amount with that released from the baking effects of the plumbing system, described above, it is possible that more than 50 gigatonnes of CO_2 could have been released to the atmosphere in a single year. The gas release would have been especially high around the start of the eruptions, when the sedimentary baking was at its most intense. As it happens, this amount is comparable to the modern rate of CO_2 addition to the atmosphere. We currently put more than 20 gigatonnes of CO_2 into the atmosphere every year from a variety of sources, mostly as a result of fuel burning but also with substantial contributions from deforestation and cement manufacture. Hundreds of gigatonnes have been released since the start of the industrial revolution. So large igneous provinces and mankind share the same ability to add hundreds of gigatonnes of a greenhouse gas to the atmosphere in just a few years. Do they therefore both share the ability to cause a mass extinction? Unfortunately we do not know the answer for the Capitanian crisis, and we will have to wait and see how our current atmospheric experiment pans out, but as we shall see in chapter 3, where we will consider the next LIP eruptive episode in Siberia, it may be that we are grossly underestimating the amount of gas produced by such volcanism.

One way of evaluating the probable climatic consequences of volcanism would be to construct a record of temperature changes during the Capitanian crisis. Such can be obtained by measuring oxygen isotope ratios in shells and skeletons because we know from the study of modern organisms that this ratio is dependent upon the water temperature in which they form. Unfortunately no detailed study has yet been done, but some fossil evidence suggests a temperature rise in the late Capitanian. Perhaps this was a response to greenhouse-gas release from the Emeishan Traps.

Another less direct clue to past temperature changes can come from the nature of the extinction losses. Many of the victims read like a roll call of the inhabitants of shallow tropical seas in the Middle Permian: fusulinaceans, reef species, corals, thick-shelled brachiopods, and crinoids. The elimination of creatures adapted to warm, shallow waters is strongly suggestive of an extinction driven by cooling, but our knowledge is biased by data that predominantly comes from the tropics. In an attempt to rectify this, Dave Bond and I have spent the past four years investigating the fossil record of the Permian rocks of Spitsbergen, an island in the Arctic Ocean, which may provide clues to the fate of marine life in the northern hemisphere. Work is in progress and unpublished, but initial results suggest a higher-latitude Capitanian crisis that was every bit as severe as that seen in lower latitudes—there may be no temperature-related factor in the mass extinction. Perhaps we should be looking for another cause?

Capitanian extinction losses have been interpreted in a different way. Matthew Clapham of the University of California–Santa Cruz and Jonathan Payne of Stanford University have been particularly keen advocates of a death-by-acidification scenario for extinctions in the oceans. They have highlighted the fact that the most extinction-prone groups had thick shells and were probably poorly buffered against the effects of acidification. Note that when applied to marine organisms, the chemical term "buffering" refers to their ability to control the pH conditions within their tissues when secreting a shell.[3] Some invertebrates, notably the bivalve and cephalopod mollusks, are especially good at this acid-base regulation today, but other groups, such as calcareous algae, are not.

The ocean's pH level is partly controlled by the effect of the dissolution of CO_2 from the atmosphere into surface waters. Any sudden increase in the atmospheric levels of this gas reduces the ocean's pH. Clearly, LIP-style volcanism is capable of doing this, and so Clapham and Payne postulated that this was the key mechanism of extinction; the loss of many thick-shelled brachiopods and calcareous algae is particularly strong evidence in support of this model. However, these organisms were not the only victims. The ammonoid cephalopods (and also several major groups of bivalves) were also devastated during the Capitanian extinction. Modern cephalopods are known to be especially good at physiological buffering and suffer little from pH fluctuations, thus the fact of their suffering during the Capitanian suggests that acidification is not the whole answer to the cause of this crisis.

It is always possible to pick and choose extinction victims to promulgate a particular extinction mechanism, but ideally all victims should be considered. As we shall see for the Permo-Triassic extinction, such selectivity by paleontologists (but not the grim reaper) has furnished ample ammunition for different extinction theories.

Work on the Capitanian extinction is ongoing, but for now we can note that the first crisis of Pangean history was severe. Most significant for terrestrial animals (i.e., the dinocephalians), it also killed many marine invertebrates while not changing the overall composition of seafloor life greatly.

Due to the precise temporal coincidence, voluminous volcanism in South China appears the most probable cause of this mass extinction, but if the whodunit seems answered, the how-did-it-do-it question is far from resolved. Possibilities include brief, intense cooling, longer-term warming, and ocean acidification. As will be seen in the following chapters, comparison with the later extinction events of Pangea indicates that ozone destruction (by volcanic halogen gases) and plant die-off (due to UV-B radiation) should also be considered as serious contenders for explaining the volcanism-extinction link at this time.

The subsequent recovery of life in the Late Permian was impressive, especially among terrestrial animals, as groups like the dicynodonts and pareiasaurs radiated rapidly to fill the landscapes vacated by the dinocephalians. Plants had suffered few significant losses, and the composition of Late Permian forests was similar to those that existed before. Marine invertebrates could also congratulate themselves on weathering a significant crisis; only the reef communities were a little slow to recover. Unfortunately for life on Earth, much worse was yet to come.

CHAPTER 3

THE KILLING SEAS
THE GREAT PERMO-TRIASSIC MASS EXTINCTION

The Capitanian was not a good time for life, but it was a relatively minor hiccup compared to the catastrophe that struck at the end of Permian, 252 million years ago. No habitat was safe during this next disaster, and all environments, from the deepest parts of the Panthalassa Ocean to the continental interiors of Pangea, saw enormous species losses. Trying to unravel the cause of this, the greatest-ever, crisis has been a major research area for geologists for more than twenty years, and thanks to these efforts some idea as to what happened is finally emerging. One aspect is clear. The culprit has long been staring us in the face: the Siberian Traps. Like the Emeishan Traps, they are a pile of basalt lava flows, only in this case much larger. Eroded remnants can still be seen over a vast swath of northwestern Siberia, and even bigger lava fields now lie buried beneath younger rocks of the West Siberian Basin. Less tangibly but more significantly, the Siberian eruptions also released vast volumes of gas and thereby created a climate of unparalleled malevolence. However, before we can understand the "how it happened" of this disaster it is important to know when and where the victims died.

WHAT DIED?

The animals inhabiting the oceans at the end of the Permian had not changed a great deal since the Capitanian crisis. The diverse inhabitants of the seabed included brachiopods, crinoids, echinoids, bryozoans, forams, ostracods, bivalves, and gastropods that could be found all over the world, and they were joined, in tropical waters, by corals and calcareous algae. Swimming in the waters above were the ammonoids, now fully recovered from their Capitanian crisis, plus nautiloids, conodonts, and fish. With a few remarkable exceptions, all of these groups were devastated at the end of the Permian. Many disappeared entirely. Thus both types of coral—the rugose and tabulate varieties—vanished, while for many other groups, such as the brachiopods, crinoids, echinoids, and bryozoans, it was a near, but not quite complete, annihilation. For many of the survivors, the intense culling had a profound influence on the evolutionary trajectory of their groups. Take the bryozoans as an example. These are small organisms that construct colonies of calcite that, in the Permian, resembled small trees and fine nets. These colony types disappeared, and today, although the bryozoans have recovered, they are very different forms, often to be found encrusting seaweed.

For some of the survivors, life was never the same again; there was to be no bounce back. Thus, the brachiopods and crinoids, which had dominated the seafloor for hundreds of millions of years prior to the extinction, never reattained their former dominance. In contrast, other groups benefited. For the mollusks, such as the bivalves and gastropods, the clearing out of seafloor life was to provide an opportunity for them to diversify and dramatically increase in abundance. Walk along almost any beach today and the shells of mollusks will dominate your finds, a legacy of the Permo-Triassic mass extinction and its killing seas, 252 million years ago. Their success is because the end of the Permian provided opportunities for the mollusks, and yet their extinction losses at that time were the most severe in their long history. The key point is that they recovered well whereas many groups did not.

The clear exception to the marine cataclysm is seen among the swimmers. The conodonts and many fish groups survived the Permo-Triassic boundary interval just about unscathed. This success is well documented for conodonts because they have a good fossil record, and it is well studied because, as we saw in chapter 2, they are the best zone fossils for dating marine rocks. For the fish though, it is a different matter. Unraveling their story is difficult due to the variable and often poor quality of fossil fish specimens. The fundamental problem is that fish, with their numerous bones, readily fall apart after death, rendering identification problematic. Fortunately for paleontologists collecting fossils, this disintegration process is minimized if the dead fish settles into quiet, oxygen-free (anoxic) settings because they are usually very tranquil and there are no scavengers to disturb carcasses. Now as we will see later in this chapter, anoxic conditions were extremely widespread in the Early Triassic but not in the Late Permian. Consequently, we have abundant, beautifully preserved Early Triassic fossils; indeed, the fossil record of fish at this time is probably the best in geological time. In contrast, Upper Permian rocks have yielded few fish, and so this period is among the least well-known intervals of fish evolution. Despite this discrepancy in the quality of the data, it appears that fish suffered very few losses at the end of the Permian. This remarkable survival is an underappreciated facet of the crisis. Fish are often the top predators in the food chain, and the fact that they survived the Permo-Triassic crisis so well indicates that at no time did the food chain collapse, despite evidence that the crisis also struck the tiniest creatures at the bottom of this chain.

The fate of planktonic groups at the end of the Permian was more akin to that of seafloor life, having major losses and changes. We know this primarily because of the radiolarians, the only planktonic group to secrete a skeleton at this time. These organisms consist of tiny and beautifully intricate lattice skeletons made of silica. Typical examples are around a tenth of a millimeter in diameter and consist of spheres nested within spheres supported by delicate radial spines that give them their name. However, their morphological variety is immense and includes tube- and rocket-shaped forms and hat-shaped types.

The weird and beautiful radiolarians are still alive today floating in the upper layers of the ocean, where they pursue a surprisingly diverse a range of lifestyles. These include grazing on other plankton, consuming organic detritus, and farming green algae. On death they rain down to the seabed to form a siliceous ooze that gradually accumulates and ultimately hardens to form a rock called radiolarian chert. Such deep-ocean cherts have been forming since start of the Cambrian, 540 million years ago, when the radiolarians first appeared. It is still forming to this day in the deepest parts of the oceans. The only interruption to this long, slow, and generally uneventful sedimentation occurred at the end of the Permian, when chert formation ceased because of a severe radiolarian extinction that was unparalleled in their history.

Yukio Isozaki of Tokyo University has studied the deep-ocean rocks in his native Japan and was the first to highlight the importance of this chert disappearance. That deep-ocean rocks can be found in Japan is due to the processes of plate tectonics. Ocean crust rarely spends more than 150 million years at the Earth's surface because it becomes denser with age as it cools. When it is sufficiently dense, it sinks (subducts) into the mantle, but fortunately for geologists, the surface veneer of oceanic sediment gets scraped off and preserved as a mangled but tangible record of ancient oceans. The deep-sea cherts of Japan are therefore a tiny and highly deformed fragment of originally widespread oceanic sediments. Other oceanic rocks are scattered around the rim of the Pacific, and they reveal much the same story of radiolarian disaster that Isozaki found in Japan. The only intriguing exception was recently discovered on the tiny island of Arrow Rocks, in New Zealand. Here many radiolarians survived the Permo-Triassic crisis, although not for long—most were gone less than a million years later. The composition of these earliest Triassic radiolarian survivors in New Zealand is unusual because they belong to types that were present before the mass extinction while only one new form appeared. So despite the fact that most of the oceans were empty of radiolarians, those that survived in the southern ocean refuge now preserved at Arrow Rocks neither recovered nor reinvaded the recently vacated habitats. There was clearly still something very wrong with the oceans after the mass extinction.

Right at the very base of the food chain, the photosynthesizing algae (phytoplankton) of the Late Permian oceans did not fossilize, because their tiny organic cells decomposed after death. Nonetheless, indirect chemical evidence of plankton populations is preserved in sedimentary organic matter and reveals a change every bit as spectacular as the demise of the radiolarians. Changes in nitrogen isotope ratios of organic matter provide an indirect way of seeing what happened to the phytoplankton. They reveal an increase of the light isotope of nitrogen, called nitrogen-14, relative to the heavy isotope nitrogen-15. This distribution suggests a shift from organic matter derived from eukaryotic green algae, the normal phytoplankton of ocean waters, to a primitive-type of photosynthesizing bacterial group known as the cyanobacteria. This inference is also supported by a concomitant increase in the abundance of organic molecules called hopanes, which are characteristic of cyanobacteria, found in sedimentary rocks. Cyanobacteria are present among modern phytoplankton populations as part of the diverse assemblages of photosynthesizing cells found in the surface waters of oceans, but their dominance during the Permo-Triassic crisis is highly unusual. Other fossil organic molecules suggest even more extraordinary changes in the kinds of plankton present at the end of the Permian.

Green sulfur bacteria are a highly unusual type of phytoplankton that photosynthesize using hydrogen sulfide rather than water, like "normal" green algae. This reaction occurs in the absence of oxygen, and so modern green sulfur bacteria are restricted to a few unusual settings where sunlight but not oxygen is available. Sulfur bacteria leave their organic fingerprint in rocks in the form of a molecule called isorenieratane, which is uniquely produced by them. This distinctive molecule has been found in the latest Permian rocks from Australia and China, indicating that anoxia must have developed in the sunlit, upper water column of the oceans—a setting that is usually well oxygenated. These findings have significance for understanding the cause of the extinction and tally with a wealth of sedimentary and geochemical evidence that is detailed later in this chapter.

So much for the oceans. What about life on land? After the extermination of the dinocephalians in the Capitanian, we left the therapsids in chapter 2 recovering, with the dicynodonts evolving into a range of plant-eating roles while the gorgonopsians were filling the large-predator niche. The pareiasaurs, a group of big, bulky reptiles with lumps and horns on their heads reminiscent of the unrelated dinocephalians, were also doing well in the herbivore role. Another diverse group at this time were the amphibians, in particular the temnospondyls, which were large aquatic animals that vaguely resembled crocodiles. More modern amphibians, known as the lissamphibians, such as frogs and toads, lay in the future although a discovery in Madagascar of a relatively advanced anuran (frog) suggests that they may already have made their first appearance in the latest Permian.

The detailed fate of all these terrestrial animals is only known from a few places, notably the Karoo region of South Africa and the foothills of the Urals in Russia, but the evidence is the same: there was a devastating extinction with loss of numerous types, including all the gorgonopsians and pareiasaurs. The only refuge from this terrestrial holocaust occurred in freshwater environments, where the temnospondyls (and probably the earliest frogs) survived relatively well.

Insects provide another clue to the fate of terrestrial life at the end of the Permian. For the most part, insect evolution has been a tale of spectacular successes and constantly expanding diversity as more and more types have been added to their impressive panoply. One of the few checks on this expansion occurred during the Permo-Triassic extinction. However, the global scale of insect extinctions is poorly known, partly because insects do not fossilize especially well (beetles with their tough wing cases are an exception to this rule), but mostly because very few paleontologists study them. One who does is Dmitry Shcherbakov of the Paleontological Institute in Moscow. His work on the Permo-Triassic insects of Russia reveals a loss of around 40% of insect families, with formerly rare groups becoming prominent in the aftermath. Intriguingly, the extinction was also followed by the northward invasion of equatorial forms, such as cockroaches and cicadas, into Siberian latitudes, a clear signal of warming. This is just one line of evidence that the extinction coincided with warming, a theme that will reappear throughout this chapter.

With significant losses among tetrapods and insects, it is not unreasonable to expect comparable changes among the plants, and this is what we find. Fortunately, we have a detailed knowledge of floral fortunes thanks to a series of studies on the spore and pollen grains found in many Permo-Triassic boundary locations. One of the first detailed looks was undertaken at Utrecht University, in the Netherlands, where Cindy Looy was studying for her PhD in the late 1990s under the supervision of Henk Visscher. I was able to contribute a little to this study by providing samples collected from outcrops in East Greenland, where I had been working with my research assistant, Richard Twitchett, on marine strata. We were able to combine our investigation of the changes in the marine realm with Cindy's spore and pollen collection and achieved a rare feat: a contemporaneous story about land and sea obtained from the same rocks.

So what did Cindy find? Initially, prior to extinction, the pollen record was dominated by a pollen grain with the long name of *Inaperturopollenites*. A gymnosperm tree produced it, and its presence indicates that there were healthy, thriving forests in the Greenland area. Near the end of the Permian the first sign of trouble was a rapid shift to shrubby vegetation dominated by spore-producing lycopsids alongside mosses and ferns. A curious and unusual feature of the lycopsid spores at this level is that many of them are stuck together in clusters of four, called tetrads. This is how they form within the plant, but when they are dispersed in the wind, the tetrads usually separate into individual spores. That this was not happening in Greenland suggests that the lycopsids were not behaving normally. In modern floras the presence of tetrads is a clear signal of stress brought on by factors such as acid rainfall, for example. Poorly formed pollen grains are also found at the same level. These usually have two air sacs that help them blow around in the wind, but in the latest Permian interval many of the pollen grains have anywhere between one and four air sacs, and they are often poorly formed; they are said to be mutant pollen. Mutants and tetrads sound like something from a horror film, and their presence at the end of the Permian indicates that plants were having a bad time.

One of the most profound effects of the Permo-Triassic extinction was to lead to the cessation of coal formation everywhere in the world. Coals are the product of lush forests growing in swamps, and perhaps, a little surprisingly, given that such environments tend to be common in the tropics today, most of the thickest coal seams of the latest Permian were to be found forming at very high southern latitudes. Thus, in Antarctica and Australia, two continents that were then in the high-latitude southern hemisphere, we find thick Late Permian coals formed from the accumulated remains of glossopterid forests. Coal formation abruptly stopped at the end of the Permian, and a new plant assemblage composed of gymnosperms, ferns, and lycopsids appeared. Many of the newcomers to these high latitudes were previously found in equatorial latitudes. The seed fern *Dicroidium* is especially diagnostic of the post-glossopterid plant assemblages, and it too was initially restricted to equatorial latitudes.[4] As we have already seen, a similar signal of warming is also seen among the insects in Siberia. Based on these observations we might expect the warm-adapted plants of the Permian to have fared better than their polar cousins, but in fact, the ability to thrive in warm conditions provided no protection at all against extinction. Plants died everywhere.

The fate of plants in the equatorial latitudes was simply a case of same story, different players. Lush forests dominated the Late Permian landscape, and the accumulation of peat swamps produced thick layers of coals. Enigmatic gigantopterid trees dominated these forests. The relationship of these trees to other plant groups is not well known, but they had many advanced features that suggest an affinity with the angiosperms (flowering plants), which appear much later. Despite their sophistication, the gigantopterids were doomed to die in the mass extinction. Triassic-type plants, especially the fernlike *Isoetes*, replaced them. However, for the most enigmatic aspect of the post-extinction flora, we have to turn to the fossil assemblages of spores and pollen because they contain abundant remains of a highly controversial spore called *Reduviasporonites*.

Reduviasporonites is a widespread spore in equatorial Permo-Triassic boundary rocks. Henk Visscher first discovered it, and he thinks it was of fungal origin. Fungi usually contribute only a very minor amount to spore and pollen assemblages. Their apparent proliferation at a time when forests were disappearing conjures up images of widespread plant decay. Alternatively, others have argued that the spore in fact derives from an order of freshwater algae known as the Zygnematales. They form the green "pond scum" that is common today in variety of ponds and lakes; therefore their abundance during the mass extinction alternatively suggests that there was an increase in such habitats at the time. Visscher, however, continues to work on this problem, and he has recently found branching chains of linked *Reduviasporonites*. These strongly resemble the chains produced by modern soil fungi that infest weakened plants. So perhaps the pond-scum hypothesis is wrong, and it is a fungal spore after all.

The death of so many plants, particularly trees, during the Permo-Triassic crisis had a profound effect on sedimentation. One consequence was a massive phase of soil erosion following the initial plant die off. Indirect evidence for this was unearthed by Mark Sephton and his colleagues (including me) a few years back. We studied the organic detritus that had been washed into the shallow seas of Tethys and found a sudden increase of a molecule called dibenzofuran, which is commonly found in soils but not in the sea. Our interpretation was that we had evidence for a pulse of soil erosion in areas that had lost their plant cover. Interestingly, the dibenzofuran peak just precedes the marine extinction, hinting at the possibility that the crisis on land may have begun just before that in the oceans.

A somewhat longer-term effect of the plant dieback is manifested as the widespread development of braided rivers at the expense of meandering varieties (a change also seen during the Capitanian extinction discussed in chapter 2). Peter Ward of the University of Washington was the first to spot this link between extinctions and river style. The explanation lies in the effects of plant roots that stabilize riverbanks and slow down the rate at which channels erode their banks. In the absence of plants, large river channels readily divide into numerous smaller, interconnected channels that are constantly changing their path. I well remember seeing this for myself while spending a day working high up on a mountainside in northern Greenland. A few miles away there was a small valley with braided channels flowing down one side of it. Around lunchtime I noticed that some new channels had started to form adjacent to the main channel belt, and by the late afternoon, the entire channel system had migrated to the other side of the valley, leaving the morning's channels completely abandoned. Greenland has very little vegetation and the majority of its rivers are of the braided variety, so I presume they are shifting around all the time, even when geologists are not watching them.

The switch from meandering to braided seen during the Permo-Triassic mass extinction can therefore be interpreted as a change from a landscape with plants to one with none. But the change in the type of river can also be caused by other factors; for example, the development of more irregular rainfall patterns favors the braiding rather than the meandering of rivers. So no explanation is ever simple, but Ward's idea is well supported by the close correspondence between plant extinction (and the consequent loss of coal formation) and the transition from meandering to braided in river styles.

EXTINCTION TIMING

The Permo-Triassic mass extinction losses are well documented for many groups and provide the key evidence for the extinction hypotheses discussed later in this chapter. Equally useful for us is that we now have very good knowledge of the chronology of the extinction, which thereby allows the course of the crisis to be understood. The duration of an extinction crisis (i.e., how quickly it happened) is particularly helpful when it comes to examining causes.

For a long time the mass extinction was thought to be a protracted crisis spread over millions of years. This was the dogma in the late 1980s, when I began my investigations of the Permian event in collaboration with Tony Hallam. At the time I had recently completed my PhD studies at the University of Birmingham, under Tony's supervision, on Jurassic fossils and strata in England and France. I had subsequently moved to the University of Leicester, but Tony and I were keen to continue our collaboration, this time on the Permo-Triassic extinction. This was something of a departure for two Jurassic devotees, but we had an idea that there was plenty that remained undiscovered about the greatest extinction and that there was an a lot of interesting work to do. Our initial fieldwork took us to an eclectic series of places: the Dolomites of northern Italy, the Rockies of Idaho, the Salt and Surghar Ranges of Pakistan, and some paddy fields in southwestern China. It soon became obvious that the received wisdom of a protracted disaster did not tally with the evidence in the field. Thus, in our first paper on the subject, Tony and I concluded that our evidence indicated "an abrupt extinction in contradiction to many previous views."[5] This statement was intended to be provocative at the time, but we found that we were pushing at an open door, and by the late 1990s a growing number of studies were in favor of a short, sharp extinction. However, there is a big fly in the ointment when it comes to evaluating the timing of the crisis: quite a few Permian species survived for a short time into the Triassic.

Most Permo-Triassic boundary rocks contain a clear extinction layer marked by the point where a large numbers of species disappear forever. However, the overlying rocks often contain a mix of both typical Permian and new Triassic fossils. This "mixed" fauna persisted into the earliest Triassic, whereupon the Permian fossils—usually called holdover taxa in the paleontological literature—disappeared. In the Italian Dolomites the mixed fauna consists of Permian brachiopods and forams, some new short-lived forms (such as the bivalve *Towapteria* and several new species of forams), plus some longer-ranging forms that become briefly abundant at this level (the brachiopod *Lingula* and the simple foram *Earlandia*, also seen after the Capitanian crisis). The significance of these fossils has long engendered debate and not a little controversy. Some have argued that the Permian holdovers are simply fossils that have been reworked from the underlying pre-extinction strata and incorporated into the younger sediments. However, this notion is easily dismissed because the mixed fauna contains new species that were not present before the first extinction. Others simply dismiss the mixed fauna as unimportant.

A recent review by Shen Shu-zheng of the Nanjing Institute of Geology and Palaeontology and twenty-one coauthors simply tried to brush the mixed fauna under the carpet when they concluded that the new species constituted a "trivial rise in diversity . . . but do not change the general overall trend of overall decreasing diversity [at this time]." In their view the mass extinction becomes a clean-cut, single event with only a few "trivial" short-lived survivors that can essentially be ignored.

But the holdover species cannot be dismissed so easily.

Clarity to this debate has come from recent intensive collecting of the Permian and Triassic fossils in South China. Sampling from numerous locations has revealed the fate of marine life in shallow seas to deep basins. Much of the hard work was undertaken by Song Haijun, who together with his Wuhan colleagues Yin Hongfu and Tong Jinnan (two giants of the Permo-Triassic research scene in China) and me, documented the fates of 537 marine species belonging to 17 major groups during the crisis. The results reveal that the mixed fauna is actually much more diverse than previously appreciated. It represents a survival phase sandwiched between *two* mass extinction events, one at the end of the Permian, which eliminated 57% of species, and one at the start of the Triassic, which resulted in 71% species extinction. Barely 40 species remained after the double-punch crisis. Thanks to much recent effort in dating volcanic ash bands in the South China sections, we also now know that the interval between the extinctions lasted about 200,000 years.

So the Permo-Triassic mass extinction as now resolved consists of two abrupt mass extinctions separated by an interval of partial recovery (fig. 3.1). If I were to play devil's advocate and argue against our own conclusions, it could be said that the extinction was in fact just one continuous phase of extinction losses spread over 200,000 years with a final coup de grâce in the Early Triassic. Declining diversity, however, does not mark the interval with the mixed fauna, called the epilogue episode by Yin Hongfu; rather, it is a time of stable diversity marked by the appearance of new species and the loss of others. In South China the epilogue episode has a stable diversity level of around 150 species. The appearance of many new species (especially among brachiopods, bivalves, conodonts, ammonoids, and ostracods) suggests that benign conditions at this time favored the appearance of these new

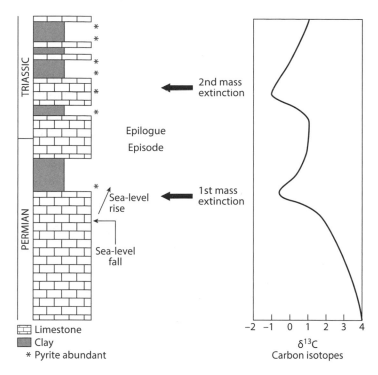

Figure 3.1. Summary of carbon isotope changes and extinction levels during the Permian-Triassic mass extinction (*right*) compared with the sedimentary beds seen at Meishan, China (*left*).

forms. These observations clearly indicate that the 200,000-year epilogue episode was not one of continued stress but rather a respite between two storms.

Interestingly, although the scale of extinction (expressed as the percentage of losses) was similar during the two extinctions, the nature of the extinction losses is somewhat different. The first extinction saw the devastation of shallow-water inhabitants: reef groups, corals, fusulinids, calcareous algae, and radiolarians all disappeared at this time. The second phase also wiped out many survivors in shallow waters, such as brachiopods, but it also removed deeper-water types, notably small forams.

Ostracods (tiny crustaceans encased in two valves) are a useful fossil group for studying the double extinction pulses because they were abundant and lived in a full range of marine habitats, from shallow to deep waters; thus it is possible to see the effects of the extinctions in different parts of the sea. During the first extinction pulse, mostly shallow-water species of ostracods were lost, whereas the second extinction also caused widespread losses, but with many shallow-water species already gone, its effect was most clearly expressed in deep water. By the end of the crisis, ostracods had been virtually wiped out except for a few survivors in very shallow waters.

The double-punch seen in the oceans also has its parallel in the fate of Permo-Triassic plants. Following the initial extinction, which saw the cessation of coal formation, there was a short-lived low-diversity period, during which shrubby vegetation and stressed lycopsids thrived, but it was quickly followed by a partial recovery. Thus, in Cindy Looy's analysis of the Greenland flora, she found the reappearance of some woodland elements such as pteridosperms (seed ferns). This interval marks the peak of plant diversity during Permo- Triassic boundary times because it contains a mix of Permian survivors and new immigrants. It is both equivalent in age to and very similar in its mix of old Permian and new Triassic species to the mixed marine fauna of the marine epilogue phase. The Greenland data also suggest that the second extinction pulse in the marine realm coincided with an equivalent extinction among terrestrial plants, which saw the loss of several important pollen types and the vegetation that produced them: *Inaperturopollenites*, from a gymnosperm tree; *Weylandites*, from a seed fern; and *Lueckisporites*, from conifers. The subsequent flora is of low diversity and lacks evidence for trees. Only the spores *Lunatisporites* and *Lundbladispora* were abundant at this time, and they indicate the presence of seed-fern shrubs and equally small lycopsids, respectively. The appearance of cycad pollen added something to the diversity and once again suggests a warming climate because this group of plants is normally restricted to low latitudes.

The two-pulsed plant extinction is also seen elsewhere in the world. In Antarctica, Permo-Triassic sedimentary rocks contain a mixed flora of the Permian and Triassic that persisted for a short time after an initial crisis the killed of the abundant glossopterids. The corystosperm *Dicroidium* is especially diagnostic of the post-glossopterid assemblages. This mixed flora persists only for around 10 meters above the last Permian coal. This thickness of rock probably accumulated over approximately 300,000 years, about the same duration as the contemporaneous epilogue episode in the sea. Following the devastation of the second extinction pulse, none of the surviving plants was more than a few tens of centimeters high. In short, it would have been possible to walk all over the world in the earliest Triassic without ever encountering a tree.

THE KILLING SEAS

Fossil evidence provides us with a clear picture of the scale of the Permo-Triassic crisis and also something about the environmental changes. The nature of the survivors, especially among plants and insects, indicates that it coincided with significant warming. Evidence from changes in plankton populations is strongly suggestive of a trend of ocean deoxygenation, but to confirm this view of climatic changes, we need to examine the sedimentary rocks.

One of the most apparent features of the Permo-Triassic marine rock outcrops is the abrupt decrease in the thickness of the rock layers from several tens of centimeters to only a millimeter or so. Such thin-bedded rocks are said to be laminated and they can, unusually, be split into paper-thin sheets (often with the help of a pen knife). Typically, marine sedimentary rocks do not show lamination because the burrowing activity of animals in the sediment disrupted the thin layers to produce beds that are much thicker. Only when life on the seafloor is inhibited in some way does this activity stop, thus preserving the lamina. One way to stop burrowing would be to kill all life in the sea, but this is unlikely; it would require 100% mass extinction, and we know that significant numbers of animals did survive in the earliest Triassic oceans. Burrow-free sediments today are most frequently encountered in anoxic settings, such as those found in the deepest parts of the Black Sea, where the absence of oxygen prevents animals from living.

Based on the simple observation of widespread laminated sediment, Tony Hallam and I suggested that the Permo-Triassic mass extinction in the ocean could be directly blamed on the spread of anoxic oceans. At the time we wrote our first paper on this topic, in 1992, we had seen evidence from Italy, China, and the western United States, and in the next few years we were to see it in many other countries.

Fortunately for geologists (but not for life in the oceans), anoxia leaves other abundant clues in sedimentary rocks. Organic fossil evidence for green sulfur bacteria has already been noted, as has the much better fish fossil record. However, one of the most distinctive clues is pyrite (which is a common iron sulfide mineral that forms in marine sediments when oxygen is absent), in particular a variety called framboidal pyrite. Framboids are raspberry-shaped clusters (up to 30 microns in size) of micron-sized pyrite crystals that form today in marine sediments precisely at the point where the last bit of oxygen disappears (fig. 3.2). It is here that the two key ingredients for pyrite formation are found together:

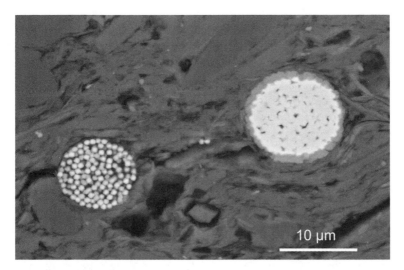

Figure 3.2. Two framboids of pyrite, consisting of clusters of tiny, bright crystals, seen at high magnification in Lower Triassic shale. Reproduced courtesy of David Bond.

hydrogen sulfide and reduced iron. As a result, numerous tiny crystals are rapidly formed and cluster together to form the framboids. Abundant framboids are a clear sign of intense anoxic conditions, but their size can also provide a further, much more subtle clue to the environmental conditions.

In some anoxic environments, hydrogen sulfide is found both within the sediment and in the waters above the sea-floor. Under these conditions, framboids form in the water column only at the sharp boundary between the anoxic and oxygenated waters and then rain down onto the seabed. This is happening today in the Black Sea, one of the few places with an anoxic lower water column.[6] Black Sea sediments are full of framboids that are all the same size—around 5 microns in diameter. They do not get any larger because the dense, little particles do not have time to grow for very long in the water before they sink to the seabed. In contrast, framboids that form in less oxygen-poor conditions, where the oxygen is absent within the sediment but still weakly present near the seabed; have time to get much larger; and consequently show a much broader range of sizes, up to 30 microns in diameter.

Using the simple principle that small framboids indicate euxinic conditions whereas a broader size range indicates slightly more oxygenated conditions, I have spent much time with my Leeds University colleagues Rob Newton and Dave Bond measuring framboid sizes in Permo-Triassic sediments from around the world. Because framboids are so small, the process requires examining rock chips with a scanning electron microscope, so the work has involved many days and weeks staring at a screen in a poorly lit, windowless room in the basement of a building at the university. This does not sound very enjoyable, but in fact the hunt for framboids on a screen is quite fun (but not yet a video game), and it is worth it because our efforts have revealed a fascinating insight into the extent and nature of anoxia in the oceans during the Permo-Triassic mass extinction.

Evidence from framboids size shows that the two phases of the Permo-Triassic crisis precisely coincide with two pulses of anoxic deposition. The first pulse was short-lived and followed by a better ventilated interval that coincides with the epilogue episode, while the second anoxic pulse was a more permanent change in marine ventilation that persisted long after the extinction. The intensity of anoxia varies with water depth, deeper-water settings generally being the most intensely anoxic. There are significant regional variations to this story. Most high-northern locations (e.g., Greenland and Spitsbergen) and some southern ones (e.g., Western Australia) show the onset of anoxia earlier than in equatorial settings and lack evidence for reventilation during the epilogue episode. Perhaps not surprisingly, these regions also lack evidence for the temporary recovery of marine life after the first

extinction pulse. In contrast, some very shallow-water locations from southern latitudes (now found in modern-day Pakistan and Tibet) avoided the first pulse of anoxia altogether.

The development of anoxia in shallow shelf seas is not a
widespread phenomenon today, but across the span of geological time it is not especially unusual. What is extraordinary
about the Permo-Triassic anoxia is its vast extent and, even
more spectacularly, the fact that the open oceans also became
anoxic. As noted above, for much of the past 500 million years,
radiolarian cherts were a typical deep-sea rock type, and their
red color indicates the presence of iron oxides. At the end of
the Permian, these red oceanic sediments disappeared, to be
replaced by black, organic-rich shales. These are full of framboids with a size distribution like those in the Black Sea, thus
indicating open-ocean euxinia. This is not the only evidence
for intense oceanic anoxia. The concentration of trace metal
concentrations in sedimentary rocks also provides a clue.

There is a fundamental chemical difference between waters
that are oxygenated and those that are not. A transition from
oxygenated to anoxic conditions causes many dissolved trace
metals to rapidly precipitate (typically as a trace constituent
of pyrite). This process is nicely illustrated by the change in
uranium concentrations during the Permo-Triassic interval.
In the normal, well-ventilated oceans, uranium is present,
dissolved in the seawater. It might sound strange that such
an element famous for its radioactivity should be present in
the sea, but it is only in tiny amounts and does not consti-
tute a health hazard. When limestones form, small amounts
of the dissolved uranium become incorporated in the calcium carbonate crystals that compose this rock type. Thus, all
limestones are a tiny bit radioactive because they have trace
amounts of uranium, usually a few parts per million. If this
were all that happened with uranium, then gradually all the
uranium in the oceans would be removed as the limestones

were buried, but this process is counterbalanced by the weathering of rocks, including limestones, on land. This process releases uranium and flushes it back into the sea. Such was the case in the Late Permian but not in the earliest Triassic, when some limestones suddenly became uranium free.

What happened to the uranium in the Early Triassic limestones? The answer has come from the study of Permo-Triassic strata in the Middle East. The limestones in this region are of immense economic importance because they include units of rock, like the Khuff Formation, which are giant gas reservoirs and so have been intensely investigated by oil industry geologists. Identifying the Permo-Triassic boundary in the Khuff is easy because it is marked by an abrupt transition to nonradioactive, uranium-free limestones. The reason for the change has to do with the onset of global ocean anoxia. The establishment of such conditions vastly increased the rate at which uranium precipitated on the ocean floor, with the result that very little uranium remained to be incorporated into shallow-water limestones like the Khuff Formation. Uranium is just one of many trace metals to show this effect (in chemical parlance, the oceans are said to be depleted); others include molybdenum, vanadium, and nickel. The depletion provides clear evidence for the extraordinary intensity of ocean anoxia around 250 million years ago.

So as geologists piece together the evidence for environmental changes during the Permo-Triassic interval, we can be confident in saying that oceans became anoxic, and fossil evidence tell us that this coincided with major warming and, of course, mass extinction. To understand just how severe the warming was requires an investigation into changes in atmospheric composition, particularly of greenhouse gases, but unlike the oceans, the clues from the rock record are very indirect ones.

ALL CHANGE IN THE ATMOSPHERE

Our best route to assessing past changes in atmospheric carbon dioxide concentrations is to analyze the carbon isotope ratios in limestones and the organic carbon found in sedimentary rocks. Most carbon atoms are carbon-12 (^{12}C), an isotope that has six protons and six neutrons; but a small amount of carbon comes in the variety carbon-13 (^{13}C), which has an extra neutron that makes it a little heavier. The proportion of the two carbon isotopes is usually given as the ratio $^{13}C/^{12}C$, which leads to a terminology in which compounds can be described as either isotopically heavy or light. Thus a rock that is isotopically heavy has a higher proportion of carbon-13 than one that is isotopically light.

Measuring carbon isotope ratios in rocks is a routine procedure for geologists and has been used for many decades. Generally most sedimentary rocks have two types of carbon: carbonate carbon, which is found in limestones and derives from the shells of animals, and organic carbon, which comes from the remains of the soft tissue of animals and plants alive when the rock accumulated. The ratios of carbon isotopes in these two types of carbon-bearing components are very different. Limestones formed in the ocean have a $^{13}C/^{12}C$ ratio close to that of the dissolved carbon in seawater, whereas organic matter is always isotopically lighter. This difference occurs because when organic matter is formed during photosynthesis, it becomes richer in carbon-12 than the original carbon dioxide from which it formed. When this reaction happens in seawater, the isotopic ratio of the carbon left over in the seawater becomes enriched in carbon-13 and thus heavier. So armed with this basic knowledge of carbon in the natural world, we can examine the isotopic history of the Permo-Triassic mass extinction to see what it shows (fig. 3.1).

Immediately before the onset of the extinctions, the latest Permian limestones have a carbon isotope ratio of +4 per mil,[7] but there is then a rapid shift to values that are 5 ‰ lighter, just as the first mass extinction struck. During the interlude phase, the ratio returns to its pre-extinction value and the limestones become isotopically heavier again. But this is a short trend because the carbon isotope ratios swing to lighter values at the second pulse of extinction. Finally, the values slowly become heavier again in the aftermath of the mass extinction. Thus, the two phases of extinction coincide with the two intervals when the carbon isotope ratios swing to lighter (more C-12 rich) values—these are called negative spikes in parlance of geologists and are a clear clue that the extinctions are tied to major changes in carbon cycling.

One measure of the huge scale of these Permo-Triassic carbon isotopic changes can be had by comparing them with the more recent history of isotope fluctuations. The past 50 million years have seen major climatic and tectonic events (an ice age and the formation of Himalayas, for example) and yet $\delta^{13}C$ values show barely a 2 ‰ change in this time. In contrast, the onset of the Permo-Triassic mass extinction saw this much change in less than 20,000 years. Clearly something big and rapid happened to carbon cycling during the Permo-Triassic transition that has had no parallel in recent geological history. Most geologists favor the hypothesis that huge volumes of isotopically light carbon were added to the atmosphere and oceans, thereby driving the $^{13}C/^{12}C$ ratio to lighter values, but just where the carbon came from and what the type of carbon was (carbon dioxide or methane or both?) remains an unresolved topic that lies at the core of many debates on the cause of the mass extinction. One main contender is the carbon dioxide released by volcanism, but it would take a lot of released carbon to cause a change the magnitude of the one observed. Alternatively, methane is a gas with very light isotopic ratios, and it would not require quite so much to be put into the atmosphere to cause such changes.

Herein lies a problem with interpreting carbon isotope values—different causes can produce similar effects. In contrast, oxygen isotopes provide an uncontroversial signal of temperature fluctuations. The tricky part is getting reliable values that have not changed since the Permian. The tiny phosphatic teeth of conodonts are the samples of choice for this work because their oxygen isotopic ratio (expressed as $^{18}O/^{16}O$) is little affected during burial and fossilization. Much careful analysis of conodonts has therefore been undertaken in recent years at the laboratory of Michael Joachimski of Erlangen University, in Germany. His most detailed results come from equatorial latitudes, which show a rapid temperature increase at the sea surface from 25°C (fairly normal for such latitudes) to 32°C (very hot!) over the interval encompassing the two extinctions and the interlude phase. As we have already seen, strong independent support for this rapid warming has already been documented from the fossil record during the interlude phase. Thus warmth-loving insects like cockroaches and cicadas migrated northward from the equator into the higher latitudes of northern Siberia at this time.

Taken together, the isotope ratios and fossils provide clear evidence for a huge environmental change, especially widespread deoxygenation of the ocean and rapid warming. All that remains is to determine how it happened, and this requires a trip to Siberia.

THE SIBERIAN TRAPS

The Siberian Traps cover a large part of northwestern Siberia and consist of hundreds of flood basalt flows stacked one on top of another to form extensive mountainous plateaus draped in coniferous forests. As with the Emeishan Traps, their original volume is hard to determine because there have been more than 250 million years during which some lavas have eroded and others have been buried. Burial has been especially important. To the west of the traps lies the West Siberian Basin, an important oil province of mostly Jurassic rocks. Deep boreholes in the region reveal that Triassic basalts underlie these younger sedimentary rocks and are probably a vast western extension of the Siberian Traps. If the buried lavas are taken into account, it is likely that the traps' original volume may have been more than 5 million cubic kilometers. Crucially, this huge volume appears to have been erupted very quickly (less than 1 million years) around the end of the Permian. Volcanic tuffs, which are produced by violent, gas-rich, explosive eruptions, were common in the initial phase of volcanism and seem to correlate with the first pulse of extinction. The second extinction pulse, in the earliest Triassic, was marked by the start of major flood basalt eruptions. The Siberian Traps are thus a very large smoking gun that was smoking at just the right time.

Attempts to understand how eruptions could cause global environmental devastation have focused on the effect of gas emissions, particularly carbon dioxide and sulfur dioxide. And herein lies the same problem, discussed previously, as with the Emeishan eruptions—the gas volumes released by large flood basalt flows are substantial, but they are not greatly different from the annual emissions due to fossil-fuel burning. So why we are not currently suffering

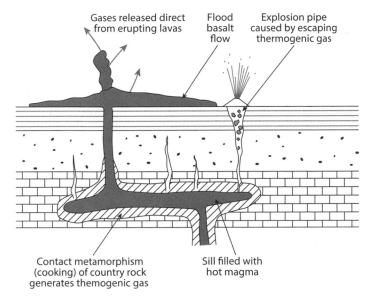

Figure 3.3. Sources of gas erupted during flood basalt volcanism.

an end-Permian-like catastrophe? There are several reasons. Either the world is much more resilient to the effects of gas emissions today than it was 250 million years ago, or there is a lag in the Earth's response system (this would imply that we just have to wait a bit longer for our Permian-like catastrophe to strike). There is a third alternative: maybe we have seriously underestimated the amount of Siberian gas erupted. In recent years most geologists have tended to favor the last alternative, and the hunt has been on to find additional sources of gas.

In one effort, Svensen's hypothesis has been invoked again (see p. 34): ascending magma heated up the rocks of the Earth's crust causing (thermogenic) gas releases (fig. 3.3). There are several reasons that make this gas-by-cooking a feasible proposition. The initial eruptions in Siberia certainly seem to have been very explosive because tuffs are widespread.

Direct evidence for the gas eruptions is also present in the form of gas-explosion pipes. These are vertical pipes of fragmented rock caused by the violent upward escape of gas—you would not want to be near one as it formed. The magmas also intruded through some crustal rocks that are likely to have generated a lot of gas, especially coal seams.

Far distant from Siberia, deep-marine rocks from Arctic Canada may provide intriguing but indirect evidence for this baking. Among the organic detritus found in the uppermost Permian rocks, there occur tiny spheres of carbon that have been likened to the char produced by the burning of coal in power stations. To get all the way to Canada, the char spheres must have been blown by the wind from the site of the Siberian eruptions.

There are, therefore, several lines of compelling evidence that the Siberian eruptions, besides releasing gas directly from the magma, also released lots of extra gas through the baking of the country rocks. There are additional indications that even more gas may have been generated by the Siberian eruptions. This evidence comes from bits of mantle rock that became trapped as solid lumps in magma and were erupted at the surface together with the lavas. Such rocks are called xenoliths, and their significance is that they provide clues to the nature of the mantle beneath the surface.

Large igneous provinces, such as the Siberian Traps, are generally thought to have come from deep within the mantle, where plumes of primitive material form and then ascend to the base of the crust whereupon they erupt at the surface.[8] This basic model has been recently much refined in a study by Stephen Sobolev and his brother Alexander, who together with other igneous geologists suggest, based on evidenced collected from xenoliths, that plumes are probably enriched in recycled ocean crust. As noted earlier, ocean crust only spends a short time (geologically speaking) at the Earth's surface before descending back into the mantle via subduction zones. The mantle thus consists of primitive material that has not undergone much melting and second-hand, recycled ocean crust. Plumes are thought to come from the deepest and most primitive parts of the mantle, but as they ascend they have the possibility to incorporate reworked oceanic crust. The latter point is important because the recycled material is much more likely to releases gases when it is heated (in geological parlance it is said to be volatile-rich). Evidence for this comes from the xenoliths in the Siberian Traps, some of which seem to be bits of former ocean crust that got incorporated into the plume and erupted at the surface.

As the plume slowly rises to the Earth's surface, the pressure decreases, thus allowing a little bit of melting and generating magma containing a lot of dissolved gas. LIP eruptions may thus have begun with gigantic gaseous burps, which have no parallel today. This is fortunate for us, but because there is nothing comparable in the modern world, it becomes rather difficult to calculate the eruption volumes. The Sobolevs have attempted a solution. Xenoliths contain tiny melt inclusions—these are bits of rock that just began to melt before refreezing and are typically less than a millimeter in size. The inclusions record the composition of the initial melt of rocks, and analyzing them reveals a very high concentration of carbon dioxide and hydrogen chloride (HCl). These gases are unlikely to have come from primitive mantle, and so the Sobolevs proposed that there was a gas-rich ocean-crust component of the plume. Scaling up their observations from the tiny melt inclusions to the scale of the plume that formed the Siberian Traps, they calculated that as much as 170×10^{12} tonnes (170,000 gigatonnes) of CO_2 (and maybe a tenth this amount of HCl) were erupted at the start of Siberian volcanism. This stupendous amount is many times larger than previous estimates of volcanic gas release and, together with gas from thermogenic heating, implies mind-bogglingly large eruptions of gas at the end of the Permian.

Intriguingly, the carbon isotope record may provide indirect evidence for the Siberian gas release. Remember that the carbon isotope ratio started getting lighter at the time of extinctions, suggesting that a lot of the lighter carbon isotope, carbon-12, was being added to the atmosphere and oceans. The isotope was unlikely to come just from the primitive mantle because the carbon from this source is not especially light. However, carbon that comes from melted and recycled ocean crust is much lighter, and so it is much more capable of reducing the $^{13}C/^{12}C$ ratio values, as we see recorded in limestones and organic matter formed at the time.

AN EXTINCTION MODEL

Based on what is known of the Permo-Triassic interval, it is possible to construct a suitably cataclysmic scenario for the course of the mass extinction (fig. 3.4). The initial blast of the Siberian eruptions in the latest Permian ejected huge amounts of carbon dioxide, sulfur dioxide, and hydrogen chloride into the atmosphere. The rapid decrease in the $^{13}C/^{12}C$ ratios of marine limestones at this time can be interpreted as evidence of a massive CO_2 input from a mantle source contaminated with recycled ocean crust. The coincident onset of rapid sea-surface temperature rise, as seen in the oxygen isotope values, is further evidence for increased amounts of greenhouse gases, and for reasons outlined below, the widespread pulse of anoxia can also be linked to this rapid climate change. All these events coincided with the first catastrophic extinction losses on land and in the oceans. Shallow marine groups were especially badly hit at this moment: reefs disappeared, along with calcareous algae, and radiolarian productivity crashed.

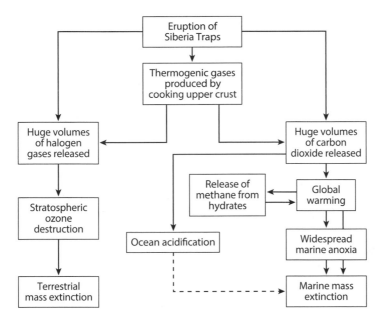

Figure 3.4. Flow chart showing the chain of events responsible for the Permo-Triassic mass extinction. The mechanism for ocean anoxia is shown in more detail in figure 3.5.

The increase in temperature may have played a large part in these extinctions because these are all groups that are the most vulnerable to hot conditions. The fact that some radiolarians were able to hang on precariously in southerly oceans for a million years or so strongly suggests that although equatorial temperatures became lethal for this temperature-sensitive group, it took a while before conditions warmed dangerously in higher latitudes. The switch to cyanobacteria-dominated phytoplankton may also be a signal of rising temperatures because this group copes much better in hotter conditions than green algae.

The vast spread of anoxic conditions at this time would have further stressed marine communities by limiting their available habitat area. Models for modern extinctions driven by habitat loss suggest that a 90% reduction in living space is required for extinction of 50% of species. For the latest Permian ocean, a figure greater than 90% for anoxic marine areas seems reasonable; there are very few regions that seem to have been entirely devoid of oxygen restriction, and these were mostly sites of very shallow water in southern Tethys. Anoxia was particularly devastating for seabed inhabitants, but it is unlikely to have been equally so for life in the upper water column, as indicated by the successful survival of fish in these environs. However, the surface waters are the setting where the temperature increase was at its most extreme, and it may have been the sole cause of death for some surface dwellers such as radiolarians. For many groups, though, temperature and anoxia stress go hand in hand because these two environmental factors act synergistically.

The oxygen requirements of organisms rise with temperature because the metabolic rate increases as life warms up. This is exceptionally unfortunate because the availability of dissolved oxygen decreases in warmer waters. Consequently, things start to get tough for most marine life once the temperature exceeds 30°C. Larger animals are particularly affected by this synergy because of the disproportionate rise in oxygen demand with increasing size—big animals need much more oxygen than small ones. The consequence of this reality is that large species are usually found in colder climes—an observation called Bergman's rule. It is thus perhaps not surprising that most of the survivors in the aftermath of the Permo-Triassic crisis were very small.[9]

Warming and anoxia may have been the key factors in the marine losses, but what about terrestrial communities? There is no hint that organisms at high latitudes were less severely affected than those at low latitudes: plant productivity crashed and coals ceased to form everywhere, indicating that this was not a regional, temperature-related crisis.

Instead, the global reach of the catastrophe implies that atmospheric changes were important. Perhaps the most likely scenario was the destruction of the stratospheric ozone shield and the bathing of plant life in harmful UV radiation. The halogen gases released by Siberian volcanism provide an obvious culprit, but to do damage they would have had to reach into the stratosphere, the layer of the atmosphere between 10 and 50 kilometers above the Earth's surface. Quiet eruptions of gases into the lower troposphere will only result in the rapid rainout of the offending gases in a matter of weeks and months; as a result, stresses such as acid rain occur locally and are not capable of doing widespread harm. In contrast, gases are much more slowly removed from the stratosphere, allowing them time to spread globally, so their effects are longer term and felt everywhere. The pyroclastic nature of the initial Siberian eruptions suggests that they were violent and so probably capable of blasting gases such as hydrogen chloride into the stratosphere. Significant ozone damage was therefore a distinct possibility during the Permo-Triassic crisis.

In addition to hydrogen chloride, other nasty halogen-bearing gases may have been implicated. Henk Visscher's group at the University of Utrecht has noted that baking of coal and salt layers in the upper Siberian crust could produce compounds like chloromethane. Such gases placed in the stratosphere can have very nasty effects because they are capable of efficient ozone destruction. Would these gases have lasted long enough to cause the plant extinctions? It is difficult to say, but some atmospheric models hint at ozone loss for periods of up to several decades after each eruption. And these may be underestimates: the Sobolevs' latest suggestions of vast eruptions of hydrogen chloride have yet to be incorporated into the atmospheric models. And other factors may also have been at play in the destruction of the world's ozone shield.

Ozone formation in our atmosphere is a very dynamic process: the gas is constantly being made and lost, but fortunately there is always a little bit in the stratosphere to shield out the Sun's damaging ultraviolet radiation. The complex atmospheric reactions involve the hydroxyl radical, which is a molecule of water minus a proton. This molecule is a "weak link" in ozone production because it is incredibly reactive and so is easily lost during other reactions. The hydroxyl radical is especially reactive with methane. Normally this is not a problem because methane occurs in trace amounts in the atmosphere, but this may not have been the case at the end of the Permian.

The shift to lighter carbon isotopes in the extinction layers could suggest that there had been a substantial release of methane from gas hydrates, icelike deposits found in marine sediments that trap methane within their crystal lattice. Methane is derived from bacterial breakdown of organic matter, and it has an exceptionally light isotopic value of −65 ‰, meaning that it is very rich in carbon-12 compared to most other carbon-bearing compounds. The release of such an isotopically light carbon source by the melting of hydrates could have driven the isotopic change. Once in the atmosphere, the methane would have contributed briefly to global warming, but perhaps its most damaging and long-lasting significance would have been its suppression of ozone levels.

Clearly there was no shortage of ozone-damaging mechanisms available for the Permo-Triassic mass extinction, but the precise correspondence between volcanism and the first phase of plant losses suggests that volcanogenic hydrogen chloride emission was the most important. Even so, the role of methane hydrate release has been a persistently popular aspect of Permo-Triassic extinction mechanisms since it was first proposed more than twenty years ago. However, the idea may have had its day. Methane hydrates form slowly beneath the seabed, and once melted by warming, they only slowly redevelop over millions of years once the oceans cool. The two negative carbon isotope spikes in the Permo-Triassic interval are separated by only 200,000 years (fig. 3.1). This is too short a time to replace the lost hydrate, and so the second light carbon spike cannot be realistically blamed on methane release. The carbon spike also occurred during an interval of continuous warming, yet cooling would be required to form more hydrates. Thus, the second spike is unlikely to record methane release, and the close similarity of the two extinction phases suggests that we are looking at the same phenomenon in both cases.

The flow chart of Permo-Triassic events seen in figure 3.4 shows a series of cause-and-effect factors that link the Siberian Traps eruptions with the global Permo-Triassic extinction crisis. Some aspects are controversial, such as the role of methane from hydrates, whereas others are on firmer ground when it comes to gases directly released from volcanism because we have the gigantic smoking gun of the Siberian Traps. The chart also serves as a blueprint for all other Pangean extinction models. Volcanic gas emissions are envisaged as the key culprits, with their consequential effects—global warming, marine anoxia, and ozone destruction—being the direct cause of death. Not all the links presented here will have operated at precisely the same time because some gases spend only a short time in the atmosphere whereas others linger. Thus, ozone destruction is a geologically instantaneous phenomenon and lasts only a few decades because halogens react rapidly in the atmosphere and once they have gone natural processes, replenish the ozone in a few years. In contrast, carbon dioxide is slowly removed from the atmosphere and causes a prolonged, cumulative warming over tens to hundreds or even thousands of years. Changes in the ocean's oxidation state are linked to this warming (see below), and they can take many centuries to achieve because it takes several thousand years for the oceans to circulate and mix.

Potentially, the contrast between the instantaneous and longer-term effects of eruptions may be detectable in the fossil record because ozone destruction and terrestrial killing may predate the temperature/anoxia–driven marine crisis even though the gases causing these changes are released simultaneously. However, the time difference may only be a few millennia, long enough on a human time scale but barely resolvable in geological strata. There is tentative evidence that the terrestrial crisis slightly predated the onset of the marine extinctions; for example, the soil erosion/plant dieback event that is seen immediately before marine extinctions in western Tethys.

Perhaps one of the most interesting new facets added to the Permo-Triassic mass extinction story in the past few years is the observation that after the first extinction phase, life on land and in the sea showed healthy signs of recovery. New species appeared during the epilogue episode, and many plants and animals that had survived the extinction in low latitudes migrated into higher latitudes as the warming continued. Thanks to these immigrants, high-latitude plant communities attained greater diversity than before the extinction—a remarkable development. This is not to say that the interlude phase marked a return to normal, pre-extinction conditions. Coal formation did not recover and ocean chert deposition had disappeared. Maybe the world was too warm for the cold-adapted siliceous organisms, or perhaps there was not enough time for their recovery before the second extinction phase struck in the earliest Triassic?

The second extinction shows most of the hallmarks of the first: huge losses in a broad range of marine and terrestrial environments and another coincidental shift of carbon isotopes to lighter values. The same chain of events can be invoked, and the same cause—a renewed burst of Siberian volcanism—can be implicated. The only difference this time was that there was no comeback for life: most groups clung on but failed to recover in the Early Triassic. This aftermath of the Permo-Triassic crisis is unique and is one of the most extraordinary intervals of geological time; all other post-extinction intervals saw recovery within tens to hundreds of thousands of years. This aspect raises several questions. Why did life fail to recover for so long? Did global environmental conditions continue to get worse, and if so, what was the cause? These, however, are questions for chapter 4.

DEOXYGENATION OF THE OCEAN

Other than the size of the Permo-Triassic mass extinction, the other extraordinary aspect of this crisis is the scale and intensity of marine anoxia. Oceanic anoxia has happened several times in the past billion years, but the example 251 million years ago was by far the most intense and is probably the only one to see nearly the entire ocean water column deoxygenated. Why did this happen? The coincidence with rapid global warming provides at least part of the answer because higher water temperatures favor anoxic conditions for several reasons. The simplest reason is the decreased solubility of oxygen as temperatures rise. This is the reason that it is not a good idea to keep a fish tank in the window—on warm, sunny days the fish will find it hard to breathe. The second, more indirect, reason is the increased rate of decay of organic matter at higher temperatures, something we are tacitly aware of because we preserve food in a cool fridge rather than a warm oven. Organic decay consumes oxygen, and so the faster it occurs, the quicker oxygen is used up.

In order to consider the significance of organic decay and oxygen consumption in the latest Permian world, we first need to understand the nature of "normal" oceans, such as those of today. Most organic matter in the oceans is formed by phytoplankton that photosynthesize in the surface waters. Small organisms called zooplankton eat the phytoplankton and produce a lot of tiny pellets of poo that form a fecal rain. As the pellets descend through the water, bacteria oxidize the organic material, causing oxygen concentrations to decline. If the fecal rain is abundant, then a point is reached, at a few hundred meters depth, where a zone of low oxygen concentration is developed. Oceanographers call this an oxygen-minimum zone (OMZ), and in a few parts of the world, beneath highly productive surface waters, very low oxygen concentrations are achieved. In these settings where the OMZ intersects the seabed, the sediments become rich in organic matter, which composes the partly degraded remains of the zooplankton pellets. Little organic matter survives decomposition below the OMZ, and the supply of oxygen is more than adequate in deep waters thanks to circulating currents. Therefore, somewhat paradoxically, oxygen levels rise in the waters beneath the OMZ, and at the present day the deepest oceans are all well ventilated.

The other main place to find organic matter in sediments is in shallow waters less than few hundred meters deep because the fecal pellets do not have far to sink from the surface waters. So in the oceans, most of the organic matter found in sedimentary rocks either accumulated in shallow waters (less than a few hundred meters deep) or in deeper settings where the oxygen depletion in the OMZ was intense.

How would the modern oceans change if they were to get a lot warmer, as happened near the end of the Permian? First, bacterial activity would increase, and so the rate of oxygen usage would go up. This would cause the OMZ to expand in thickness and become more intensely anoxic at its core, resulting in more oxygen-poor waters. Other more profound changes would probably happen. Radiolarians would become extinct, suggesting a crisis among plankton grazers that include zooplankton. This could have happened in the past because of the transition from green algae to cyanobacterial-dominated plankton, which is seen in the fossil record of organic molecules. Zooplankton are not capable of grazing tiny cyanobacterial cells. In an ocean without zooplankton, there would be no tiny, zooplankton fecal pellets. A poo-free ocean does not sound too bad, but it would be a catastrophe because this is how dead planktonic material gets to the seabed. The absence of a mechanism for packaging organic matter into little pellets would mean that the dead cyanobacteria cells would just hang around, floating and decomposing in the water column, greatly increasing the rate of oxygen usage and so causing the waters to become oxygen free.

All told, the latest Permian temperature rise might have killed radiolarians (and zooplankton), caused a switch from green algae to cyanobacteria, weakened the plankton food chain, decreased the rates at which organic matter sank, and increased the rate of bacterial decay, thereby providing a multiplicity of factors that all led to intensification of water column oxygen consumption (fig. 3.5).

A second reason that warming may cause deoxygenation concerns the nature of ocean circulation. When originally advocating a cause for Permo-Triassic ocean anoxia in the mid-1990s, I deployed what I then thought were well-established ideas concerning ocean circulation. These related to the density differences in waters of different temperature. Generation of cold surface waters in polar waters increases their density, allowing them to sink and form a deep-ocean current system that is balanced by the poleward flow of warmer, less dense, tropical surface currents. Now it would seem reasonable to argue that the rate of this mixing depends on the rate of generation of cold water. Therefore a warmer world, with

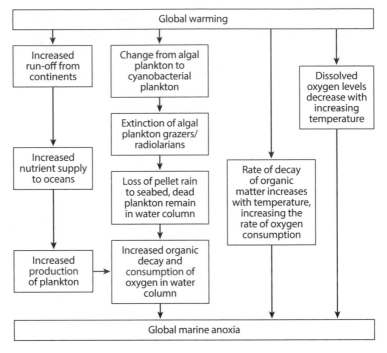

Figure 3.5. Deoxygenation of ocean waters.

warmer poles, should see a decline in ocean circulation, or so I reasoned. However, the 1990s also saw the first generation of coupled ocean-climate computer models that indicated that although decreased ocean circulation did correspond to warming, it was caused by increased river runoff in higher latitudes. This resultant freshening of surface waters decreases the density and effectively reduces polar sinking rates. So it may be salinity rather than temperature that is of primary importance, but the effect on circulation is the same. All well and good, but science is always moving on and the latest ideas are now very different.

Current research on ocean circulation ascribes great importance to wind-driven currents; for example, the strongest ocean current in the world today flows around Antarctica and is driven by the powerful winds at these latitudes. So, will a warmer world be more or less windy? Atmospheric circulation models suggest that wind strength is mainly a response to the thermal contrast in the middle of the atmosphere, and that difference increases with global warming. So a warmer world will be windier, and the oceans will be better mixed. This view is counter to that prevailing only a few years ago and highlights how fast our views of the ocean-climate system have changed in just a few years.[10]

A partial clue to the nature of ocean circulation in a warm world comes from the recent geological history of the Mediterranean. Today this warm sea has a circulation that is driven by intense evaporation because the climate is generally dry and warm (making it a favorite holiday destination for many northern Europeans). As the sea surface evaporates, dense, saline waters are formed, which sink to the seabed and flow out of the Mediterranean through the Straits of Gibraltar. This flow is counterbalanced by a surface inflow of less-salty water from the Atlantic. The downwelling of surface waters removes nutrients needed by plankton, with the result that Mediterranean waters are some of the least productive but best oxygenated in the world. However, only a few thousand years ago, not long before the first Egyptian pyramids were built, the eastern Mediterranean was more productive and large parts were anoxic. The cause was a greater inflow of freshwater and nutrients from the Nile, which decreased downwelling and stimulated increased plankton productivity, thereby intensifying oxygen usage. This change emphasizes the subtlety of the controls on ocean oxygenation. In the case of the eastern Mediterranean, it is the amount of rain falling on the Ethiopian highlands that feed the Nile that is the key driver, but the ultimate control is once again warmer climates and the associated higher humidity.

The whole-ocean anoxia of the Permo-Triassic mass extinction is an extraordinary phenomenon that has attracted the attention of many computer modelers who have been keen to understand the conditions required to produce such a state. Initial efforts ten to fifteen years ago used 3D ocean models with prescribed atmospheric conditions ("prescribed" means the modeler has to decide what these conditions are before the model is run). Initial results only achieved transient and/or regional anoxic conditions but could not replicate the widespread and persistent ocean anoxia seen. By the end of the 1990s some modelers were claiming that end-Permian ocean anoxia was highly unlikely because their models did not reproduce it. As computer power has increased, the modeling efforts have become more sophisticated and Earth system models of intermediate complexity (EMICs) are now used. In these programs the atmospheric conditions are modeled, so there is no need to guess them. Early EMIC runs looked at the effects of increasing atmospheric CO_2 concentrations to 18 times the preindustrial values of the modern world. The results unsurprisingly showed that adding a lot of this greenhouse gas to the atmosphere caused a strong increase in sea-surface temperature, but there was also a much more vigorously circulated ocean that remained fully oxygenated. More recent and more sophisticated EMIC efforts have not been able to change these conclusions: substantial warming only serves to increases ocean mixing and thereby maintains well-oxygenated oceans. Thus, the latest (2012) study of Cornelia and Arne Winguth concludes with the categoric but demonstrably wrong statement that "the deep Panthalassa remained ventilated."

Computer models are like sheets of toilet paper: they are very useful on the day you want to use them, but generally you do not reuse them the next day. In other words, they are constantly being replaced as new ideas and new parameters are developed and used in increasingly sophisticated models. When they fail to model what is known to be true from the physical evidence, then they need to be changed. In 2008 Katja Meyer (then at Penn State) and her colleagues yet again modeled the effects of high-atmospheric CO_2 concentrations, but this time they also evaluated the significance of increased nutrient input to the oceans. As a consequence, they were able to reproduce reasonably intense anoxia within some parts of the Permo-Triassic oceans: not quite the real situation at the end of the Permian but a much better result. Their conclusion, that nutrient supply is the key factor, has echoes in the recent behavior of the Mediterranean. Cogently identifying the known unknowns of their model, they concluded that "further work is necessary [in order] to constrain the nature of the biological pump during this [mass extinction] interval." The "biological pump" refers to the mechanism by which surface-water organic matter is transported to the deep, and it harks back to the point I noted above: that the demise of zooplankton and their fecal pellets could have been a powerful feedback mechanism for enhancing anoxia during the mass extinction.

Computer models inevitably have as their starting point an understanding of how present-day oceans work, but their relevance to understanding extraordinary intervals is limited by our inability to know a priori what was going on. The Permo-Triassic oceans were clearly fundamentally different from those today, and the key reasons are likely to have been the lack of any fecal rain, a switch to cyanobacterial plankton, and the vast increase in oxygen consumption by bacteria in warmer waters. None of these phenomena have been used in existing computer models.

OTHER SYNERGIES

So we are at the point where the killing seas of the Permo-Triassic mass extinction can be explained by the stressful synergy of oxygen loss and temperature increase. The responsible gas emissions were mostly likely carbon dioxide from the Siberian eruptions, possibly with some help from methane release from hydrates. However, an increase in atmospheric CO_2 can also lead to increased concentrations of this acidic gas in ocean waters, raising the specter of a third synergy, ocean acidification, as a factor in the mass extinction.

Ezat Heydari of Jackson State University was the first to look for field evidence for ocean acidification. He focused most of his efforts on some Permo-Triassic boundary sections of Iran where a thin bed of calcareous mudstone is interlayered in a succession of limestone beds. Although mudstones are the most common of all sedimentary rocks, Heydari attributed special significance to the boundary mudstone by claiming that its carbonate-poor nature was caused by intense dissolution in an acidic ocean. On its own, a mudstone bed is not very compelling evidence for acid waters; it could just as easily have been formed by a slight increase in rainfall resulting in more mud being washed into the sea. However, there is also evidence from the nature of the extinctions. Thus, the heavy losses suffered by animals with thick shells, such as brachiopods and corals, is frequently offered as proof of death by acidification for the reason that it is hard for animals to produce a calcium carbonate shell if the waters are a little acidic.

As to the cause of acidification, one suggestion proposes that upwelling of deep anoxic waters into shallow waters would have brought dissolved hydrogen sulfide gases into surface waters, where they would react to produce sulfuric acid—an acid-bath death. In this scenario I cannot help but think that the hydrogen sulfide itself would probably do the killing, because it is a lethal gas for all animals, but the main problem with this idea is the timing. Deep-ocean hydrogen sulfide levels developed coincidentally with the first-phase extinction, allowing very little time for much hydrogen sulfide to build up in deep waters. In fact, most other proponents of the acidification kill mechanism generally favor increased CO_2 concentrations as the cause of the reduced pH.

How lethal can ocean acidification be? Was it part of the litany of end-Permian woes, or was it a peripheral (or even irrelevant) stress at best? And how good is the evidence for this kill factor? Is the preferential loss of thick-shelled organisms really suggestive of acid-driven extinction? There were certainly severe losses among these groups, but they would also have been susceptible to the temperature increase and oxygen decrease as well, two factors that are known to have actually happened during the mass extinction. When looking at the relative fates of marine organisms in a crisis, it is important to take a broad view and not just pick out bits and pieces of evidence that fit a specific idea. The key observation is that the crisis was devastating for nearly all life, including groups like the radiolarians and worms,[11] which should have weathered acidification stress with ease because they do not have a calcareous skeleton prone to acid dissolution. Similarly, the agglutinating forams suffered their worst ever crisis at the end of the Permian, and yet this is the acid-tolerant group par excellence in modern seas. They can be found today thriving adjacent to carbon-dioxide-venting volcanic sites in some of the most strongly acidic waters found anywhere in the oceans. Finally, mollusks faired better than brachiopods and yet the former have shells made of aragonite, which has a significantly greater solubility than the calcitic shells of the latter. If acidification was an important cause of extinction, then the brachiopods should have faired better than mollusks. A role for acidification on the Permo-Triassic extinction story is at best unclear, but it appears unlikely.

Before leaving the acidification discussion behind, it is worth taking a look at the nature of modern oceans and the debate about their currently rising acidity levels. This discourse is relevant to mass extinction studies because it may help us to understand acidification effects and, conversely, geologists usually have an eye on the relevance of their conclusions to near-future oceans in crisis. Of course, the advantage of studying the present day is that direct measurements of water acidity levels can be made. These reveal that pH levels have been dropping in the past decades as a consequence of increased atmospheric carbon dioxide (and we all know who is responsible for that) with possible concerning consequences for the ocean's shellfish. Huge research programs are underway to assess this observation.

The findings so far have been ambiguous. The pH changes predicted for the next century are likely to increase the rate of shell dissolution, but only by a small amount; and for most organisms, which have lifespans measured only in a few years, this change will be unimportant. For example, if a bivalve lives 10 years, its shell will not dissolve much in this time. And yet, long after it is dead, the shell will continue to dissolve and start to get a little thin and fragile, but the bivalve will be past caring by then. The only seafloor community where the longevity of shells really matters is in reefs, which are constructed of the shells of animals both living and dead. Potentially a slight increase in dissolution may weaken the reefs' construction. However, the main concern is really whether increased acidity will make it harder to secrete shells in the first place. For most organisms, shell formation does not occur in open seawater but within their own body tissue, where they can locally elevate the alkalinity in their fluids to facilitate calcium carbonate precipitation. Only a few simple groups, like calcareous algae, rely on the ambient conditions for the formation of their shells.

Many studies are now coming to appreciate that modern sea life is facing a range of threats, but acidification appears to be, for many species, a relatively minor one. A recent paper by a team based at Bangor University in Wales succinctly stated the case in the title of their paper: "Ocean warming, rather than acidification, reduces shell strength in a commercial shellfish species during food limitation."[12] I suspect this could have been a pertinent headline at the end of the Permian, too.

Carbon dioxide is a very troublesome gas, and there is a final synergy that must be considered: hypercapnia, the stress experienced in animal tissue caused by carbon dioxide build-up in body fluids. We experience it as a stitch after vigorous exercise, but fortunately it is rarely more than a few minutes of inconvenience. However, for aquatic organisms it can be more serious because they are bathed in water that can dissolve significant levels of carbon dioxide that can in turn retard their metabolism.

Andy Knoll of Harvard University has developed these observations to suggest that hypercapnia was a direct cause of extinctions at the end of the Permian. Once again, his claim comes from analyzing the selectivity of marine losses. Grouping organisms into three categories based on their perceived ability showed apparently distinct differences in their generic extinction losses (table 3.1). Group 1 organisms are those with thick carbonate skeletons, slow metabolic rates, and a poor ability to respond to carbon dioxide fluctuations. Group 2 animals are thinner shelled and considered to be somewhat more responsive, and Group 3 contains animals without carbonate shells, that are predicted to be untroubled by hypercapnia.

Unfortunately for the hypercapnia hypothesis, this data set fails to stand up to much scrutiny. Even a cursory glance at

Table 3.1. Generic extinction percentages proposed for various animals during the Permo-Triassic mass extinction.

Animals grouped according to their perceived ability to withstand hypercapnia.					
Group 1		*Group 2*		*Group 3*	
Rugose corals	100%	Gastropods	42%	Polychaete worms	0%
Bryozoans	35%	Bivalves	55%	Holothurians	0%
Brachiopods	86%	Ammonoids	37%	Conodonts	33%
		Nautiloids	38%		
		Ostracods	77%		
		Malacostraca	40%		

the table shows discrepancies: the Group 3 conodonts have extinction rates more akin to those in Group 2, and the Group 1 bryozoans, with their low extinction rates, should surely also be in Group 2. More fundamentally, the fossil record of holothurians, polychaete worms, and the malacostracan crustaceans is very poorly known and it is misleading to include them in the analysis. The 0% extinction claims for worms is especially strange. It is not supported by the observed loss of nearly all burrows at the end of the Permian; such trace fossils, as they are called, provide an insight into the activities of soft-bodied organisms in the sediment. Many different types of trace fossils disappear at the end of the Permian, including an impressive feeding trace called *Zoophycus*. They do not reappear for many millions of years, during which new worms had to evolve to replicate this lost burrowing type. Equally significant are the many groups not considered in the analysis, especially organisms with siliceous skeletons like sponges and radiolarians. On the basis of their mineralogy, radiolarians should be placed in Group 3, and yet their extinction rates suggest Group 1. Hypercapnia is also a potential stress for modern marine life as the levels of dissolved carbon dioxide increase, but most studies rank it as a relatively minor concern compared with warming, decreased oxygenation, and (potentially) acidification. This is likely to have been so in the past as well.

The scenario for the Permo-Triassic mass extinction outlined in figure 3.4 summarizes the best understanding we currently have about the cause-and-effect cascade triggered by the giant initial burst of the Siberian Traps. It contains several direct kill mechanisms such as anoxia and temperature but eschews other proposed, popular mechanisms, such as acidification and hypercapnia. It has long been the case that Permo-Triassic extinction studies have tried to accumulate as many different ways of killing life as possible. Such is the scale of the losses that many geologists have wanted to throw as many kill mechanisms into the pot as possible. In his valuable summary of the field in 1993, Doug Erwin memorably likened the job of evaluating the various extinction hypotheses to the task of Agatha Christie's famous detective Hercule Poirot in the *Murder on the Orient Express*. In the end Poirot concludes that all the passengers committed the train-based murder because all had various unconnected motives. In essence, Erwin was suggesting that the Permo-Triassic mass extinction to be the perfect storm of unfortunate coincidences.

The "Orient Express" viewpoint has proved popular ever since, but it is a most unfortunate one because it has discouraged attempts to critically evaluate proposals. Instead, new suspects are continually added to an ever-increasing list. Thus, Mike Benton and Andrew Newell's recent evaluation of terrestrial extinctions concluded with a complex, multi-factor extinction scenario: "Tetrapods may have succumbed primarily to acid rain, mass wasting, and aridification [while] plants may have been more affected by the sudden effects of heating and wildfires." Andy Knoll's recent summary of marine extinctions is equally complex: "Multiple killers were undoubtedly at work [such as] hypercapnia [and] hypoxia [while] sulfide generation and temperature increase likely operated synergistically with elevated P_{CO_2} to cause selective extinction." These are not so much the four horsemen of the apocalypse as an entire cavalry brigade. Uneasy with invoking abundant coincidences, Knoll suggests that his "new focus is not a resurrected 'Murder on the Orient Express' scenario in which unrelated bad things happen by chance but rather the coordinated imposition of physical perturbations" (p. 308). Well, maybe, but it would be a scientific improvement to rank the severity of all these multiple factors.

Rather than swamping the subject with a plethora of causes that require the Permo-Triassic crisis to be the most anti-serendipitous interval of all time I prefer adopting the strategy of another detective, Sarah Lund the central character in the Danish TV crime drama *Forbrydelsen* (remade as *The Killing* and set in Seattle). In the first series Lund is confronted with a murder and a list of plausible suspects that grows longer with every episode. Avoiding Poirot's frankly rather lazy "everyone did it" conclusion, Lund eventually discovers the singular killer, who was someone rather obvious from an early stage. With this in mind, I propose a "Forbrydelsen" approach to Permo-Triassic extinction causes. We have a single, big smoking gun and need only acknowledge that sometimes volcanism is so vast in scale that its effects, especially warming (and its handmaiden ocean anoxia) and ozone destruction, can be catastrophic.

CHAPTER 4

TROUBLED TIMES IN THE TRIASSIC

If you were able to stroll along an equatorial shore in the Scythian—the alternative name for the Early Triassic—looking for shells, you would be sadly disappointed by the lack of variety. Choosing to wade into the sea to look for richer pickings, it is unlikely you would get very far because scalding temperatures would drive you back onto the beach. Disappointed, you might head inland away from the sea, but it would be a lonely stroll. Sparse scrubby vegetation and an absence of animals would be your lot. Wandering thousands of miles across the equatorial belt of central Pangea, you would see nothing larger than the occasional scorpions. However, this untrodden solitude would not be a Sahara-like desert but an environment without parallel today. The humidity and heat would be overwhelming, and frequent storms would create spectacular, short-lived flash floods. Heavy rainfall and warmth is normally associated with luxuriant forests and diverse life, but neither would be found in the aftermath of the Permo-Triassic extinction. Such a lifeless state could be blamed on enormous scale of the preceding losses, but other mass extinctions were typically followed by recovery within hundreds of thousands of years.

The Scythian low in diversity lasted for 5 million years, and only in the Middle Triassic do we see ecosystems beginning to thrive once again. Indeed, for many groups that had survived the great extinction 252 million years ago, conditions only got worse. The idiom "out of the frying pan and into the fire" is not only appropriate for the Early Triassic, it comes close to being literally true.

Our knowledge of the Early Triassic world has changed fundamentally in the past few years as the result of an intensive series of research programs that can be traced to the pioneering work of Dave Bottjer and several generations of his research students at the University of Southern California. In a more modest way I also instigated this burgeoning research because, in the early 1990s, one of my own research students, Richard Twitchett, began investigating Scythian fossils found in the Dolomite Mountains of northern Italy. Prior to this, the few species to be found had attracted little attention, but subsequent research has shown these fossils to be remarkably strange. The same few species, mostly bivalves and gastropods, turn up everywhere in the world, making it the most widespread fauna there has ever been. Equally fascinating, the fossils are frequently of small size; Richard has called it a Lilliputian fauna, after Jonathon Swift's race of tiny people in *Gulliver's Travels*.

Also unusual, Lower Triassic limestones often contain mound-shaped structures called stromatolites. These can still be found forming today in a handful of locations where cyanobacteria on the seabed trap sediment and gradually build up layers to form domes around a meter or so in height. In some cases the Triassic mounds lack the layering and instead show a blotchier internal appearance—these are thrombolites. Both of these types of microbial buildup, as they are called, appeared in the aftermath of the end-Permian mass extinction, and they continued to proliferate in the earliest

Triassic. Their abundance at this time was decidedly anachronistic. Such microbe-built structures were very common for several billions of years in the Precambrian oceans, but the rise of animals at the start of the Cambrian, 540 million years ago, saw them become extremely rare. Their resurrection in the Early Triassic is therefore very odd. Do the stromatolites and thrombolites record the return to Precambrian-like ocean chemistry or some other strange oceanic conditions? The most probable explanation is that the Triassic oceans had become saturated with carbonate, like Precambrian ones, ensuring that it took little effort for microbes to cause carbonate to precipitate.

THE SURVIVORS

Despite the general impoverishment of Early Triassic marine life, there are a few success stories to be found, especially among those animals that swam in the water column. Most fish had easily survived the Permo-Triassic crisis, and their fortunes got even better at the start of the Triassic, when several new types began to appear. These included the neopterygians, the group that includes nearly all the modern ray-finned fish, although true modern-style fish, called teleosts, did not appear until later in the Triassic. Also new to the scene were the neoselachian elasmobranchs, which include all modern chondrichthyians, known to us as the sharks. The Early Triassic also saw the appearance of strange, short-lived evolutionary innovations including the coelacanth *Rebellatrix*, which was, unusually for these normally sluggish creatures, adapted to high-speed, sustained swimming. One of the most successful new fishes in the Early Triassic was the genus *Saurichthys*. Its highly elongate morphology resembles that of modern belone needlefish, but this was the first time a form of this shape had evolved, and it proved to be highly successful. Numerous species are known from both marine and freshwater locations, and its distribution in the Early Triassic was remarkably widespread, from pole to pole. Significantly though, *Saurichthys*, like most other fish, was not found in equatorial Tethys in the Early Triassic and yet, come the Middle Triassic, it became common in this region. Such absence from the tropics is seen among many fish and provides an intriguing clue to the world's climate at this time—but more on this later.

Spectacular radiation like that shown by the fish usually occurs after a mass extinction. The rise of the mammals after the death of the dinosaurs is the quintessential example, but curiously, the fishes were not rebounding, because they had not suffered any major losses during the Permo-Triassic crisis; their evolutionary success was not a case of refilling vacated environments that had lost their previous incumbents. Presumably it was the changed conditions of the Early Triassic oceans that fired the starting pistol for an evolutionary fish race. The fish really are a remarkable group, and having successfully ignored a catastrophic mass extinction that devastated most marine life, they then began radiating like a huge group of revelers who turn up at the end of a funeral and starts a riotous party.

Ammonoids were also a significant component of marine diversity in the Early Triassic, but not because they had weathered the extinction especially well—on the contrary, they were nearly wiped out—but, rather, because they bounced back quickly. In many ways this recovery was predictable. Throughout their long history, ammonoids had extraordinarily rapid evolutionary rates, faster than those of any other animals. New ammonoid species constantly appeared and went extinct, while in the background, "dull" creatures like bivalves made do with the same few species, which plodded on unchanged for millions of years. It has been calculated that within a million years of the Permo-Triassic extinction, hundreds of new species had appeared. The ammonoids, like the fish, swam around in the water column, and so their success indicates that this part of the marine habitat was fine after the mass extinction. Furthermore, it indicates that despite the devastation and losses during the crisis, a well-developed food chain having large carnivores such as fish at its apex continued to persist.

The success of the ammonoids is a tangible record of recovery in the Triassic seas, but it was not quite what it seems. Simple counts of new species and genera clearly show a spectacular rebound, but many of the new forms that appeared after the mass extinction were all rather similar in appearance. They tended to be simple, smooth forms. Therefore, one of the effects of the Permo-Triassic mass extinction losses was to remove a variety of different ammonoid shells. In biological parlance, morphological diversity is called disparity, and so a more nuanced description of ammonoid recovery would state that although pre-extinction species diversity levels were attained very quickly, the recovery of disparity was slower. Disparity levels were not back to Permian standards until around 3 to 4 million years after the crisis. This was about the same time that other marine groups were starting to recover. It may have been that ecosystems were simpler in the earliest Triassic and so the usual evolutionary exuberance of ammonoids was somehow muted: they were constrained to get by with the few morphological types that could survive at that time. However, if this scenario is correct, then it does not apply to fish, which, as we have seen, achieved both high disparity and diversity in the Early Triassic.

In contrast to life in the water column, the inhabitants of the seafloor remained moribund. Most groups showed few signs of diversification until the Middle Triassic; for example, it was more than 10 million years before a new coral group, the scleractinians, appeared and began to construct small, simple reefs.

The situation with terrestrial communities was much the same. Plant communities were simple and composed of herbaceous lycopsids, pteridosperms, and mosses—a remarkably archaic flora that is reminiscent of the plants found in the wetlands of the Carboniferous more than 60 million years earlier. However, unlike the Carboniferous Period, which is famous for coal formation, no coal swamps were to be found anywhere in the world in the 10 million years after the Permo-Triassic mass extinction, a sure signal that plant communities were not very luxuriant. Equally strange, the Early Triassic plants were not very big, rarely growing above knee height, whereas their Carboniferous forebears had included many giant trees.

The only evolutionary innovation among plants at this time was in the southern polar latitudes of Gondwana, where two new groups of seed plants, the cycads and bennettitaleans, proliferated. These would go on to spread around the world in the Middle Triassic and become the most significant plants of this time. The former plants still survive today, and their stout stems and crown of long fronds are popular ornamental plants around the world. I have two luxuriant examples in my office, which are admittedly too big because students in tutorials often have to sit peering out from among their leaves. The bennettitaleans, on the other hand, are now long gone, and because they are extinct, their relationship with other plants is unclear. They were rather like cycads with some similarities to more modern flowering plants.

As with plants, terrestrial animal communities remained low in diversity for millions of years after the Permo-Triassic crisis. *Lystrosaurus* was a pig-sized dicynodont that rose to prominence immediately following the extinction episode.

For the first million years or so, these low-diversity communities had many vacant ecological niches; there were no small herbivores or large predators, for example. The absence of the latter perhaps explains why *Lystrosaurus* was so abundant and widespread. In a world without enemies, it went forth and multiplied. Certainly, based on its appearance alone, it is not readily apparent why this animal was so successful. It was a typical dicynodont, having a stout barrel-like body and blunt face armed with strong, short tusks. These may have been useful for digging out tough vegetation, and it is possible that its hippo-like form was adapted to a semiaquatic lifestyle. Rivers and lakes were certainly one of the few terrestrial environments to witness any evolutionary success in the Early Triassic because amphibians—a group whose major heyday had been earlier, in the Carboniferous Period—underwent their last major diversification in the Early Triassic. As noted in chapter 3, the temnospondyl amphibians survived the Permo-Triassic crisis reasonably well,[13] and it was followed by their rapid and spectacular radiation. For the most part, it consisted of the evolution of rather compressed, crocodile-like body forms that ranged in size from a few centimeters up to a few meters.

Finally, having left the greatest Scythian success story for last, we turn our attention to the insects. The Permo-Triassic mass extinction was the nadir in the fortunes of insects, but the Triassic was to be their heyday, the time when the modern insect groups, familiar to us today, began a stupendously successful radiation. The explosive diversification saw the appearance of many groups of Coleoptera (beetles) along with the Diptera (true flies), the Hymenoptera (which includes the wasps and bees although such social insects did not appear until after the Triassic), and the Heteroptera (true bugs, such as the common visitors to my garden, the shield bugs). This success began almost immediately after the extinction; for example, the oldest locust fossils occur in sediments interlayered with the lavas of the Siberian Traps, while many new aquatic types (including mayflies and many beetles) are found in freshwater sediments from very start of the Triassic. The only insects not to do well at this time were the xylophagous (wood-eating) beetles, for the obvious reason that woody trees were absent from the flora of the Early Triassic. An indirect measure of insects' success at this time comes from the radiation of the cynodonts that fed on them. These small animals first occur in the latest Permian, and having survived the extinction, they began a diversification in the Early Triassic that was to give rise to the mammals later on.

THE GREATEST GREENHOUSE

Life's successes in the Early Triassic were decidedly few and far between. In the oceans the fish and ammonoids were clearly doing well, and on land the river and lake communities with their new groups of amphibians and insects were thriving. But what was holding back everything else? A possible answer for the oceans is found in the sedimentary rocks. The anoxic black shales that developed at the time of the extinction continued to be widespread, thereby restricting seafloor life to narrow habitable zones around the shoreline, where oxygenated conditions were to be found. Some have argued that low oxygen levels in the atmosphere may also have inhibited terrestrial life; however, there is little evidence to support this notion and no viable mechanism to cause a dramatic drop in atmospheric oxygen levels. The radiation of insects, including many large forms, would also have been unlikely if atmospheric oxygen had been low because they rely on passive diffusion and high ambient oxygen levels for respiration. Charcoal is also common at this time, indicating the frequent burning of plant material, which is only possible if atmospheric oxygen concentrations stay above 16%, a level that is more than adequate for life on land.

Rather than poor oxygenation, the key factor in explaining the unusual conditions of the Early Triassic is probably extremely high global temperatures. Greg Retallack was the first to stress this aspect, based on his analysis of fossil soils, and he memorably called the interval a post-apocalyptic greenhouse. As noted in chapter 3, Michael Joachimski has shown that sea-surface temperatures had risen rapidly during the course of the twin-pulse mass extinction. Further work by Michael, myself, and Sun Yadong, together with colleagues from Wuhan, suggests that this was just the start of a spectacular ocean temperature rise of 20°C. Our evidence is two fold: fossils and geochemistry. The latter consists of oxygen isotope values derived from conodont animals. Like other fish groups, they survived the Permo-Triassic mass extinction very well, and like most other survivors, they became very small in the aftermath. Unfortunately, this means that their little phosphatic teeth—the only bit of the animal that fossilizes and the bit that paleontologists collect by extracting from rocks—became even smaller (typically less than 500 microns, or 0.5 millimeters, in length). This is unfortunate for geochemical analysis because it means that a lot of conodonts are needed for each analytical sample. Fortunately, Yadong is capable of the hard work required, not to mention being skilled at very fiddly analysis. By dissolving two tons of limestone from South China and spending months picking through the residue with a fine paintbrush, he was able to collect and identify 15,000 conodonts. Analysis of the oxygen isotope variations of this collection revealed a temperature curve for sea-surface temperatures that began at 20°C shortly before the mass extinction and peaked at 40°C in the Early Triassic. At no point did temperatures drop below 32°C in the 5 million years after the extinction. These values are exceptionally hot; in fact, the highest ever recorded. For comparison, modern equatorial sea surface temperatures typically average around 28°C and never exceed 30°C. Early Triassic seawater temperatures were therefore extraordinarily hot, with peak temperatures the same as you would find in a bowl of very hot soup.

Without supporting evidence, our calculated temperatures for the Early Triassic could be dismissed as a problem with the analyses or the samples. Perhaps the oxygen isotope signal had been altered during the burial of the conodonts, making it unreliable. However, a superhot climate also explains the extraordinary nature of the fossil occurrences at this time; for example, Bergman's rule readily explains the dominance of small animals in the earliest Triassic. This is one of the most commonly observed "rules" in biology, and it notes that as temperatures rise, animals get smaller. There are several reasons why this happens. As temperature increases so does growth rate, which causes adult body size to be reached quicker. Hotter conditions also increase the rate of juvenile mortality, which means that the fossil record, in effect a census of dead things, becomes dominated by a higher proportion of small individuals. More dead juveniles and smaller adults in hotter conditions all combine to produce a lot of small fossils.

Animals in a hot world live by the strategy live fast and die young. However, when the temperature gets too hot, even short lives become difficult, because of a fundamental problem. As temperatures approach 40°C, proteins begin to unravel, a process called denaturation, which is immensely harmful to life. It can be countered by the production of heat-shock proteins, but that requires metabolic energy and can only provide a temporary respite. Aerobic performance also increases with energy, which is not a bad thing in itself, but at around 40°C, the mitochondria struggle to keep up with a cell's energy demands. We know this from our own experience of intense lethargy when we have a fever, which causes our body temperature to increase above its normal 37°C value. One strategy for animals to avoid some of these problems is to have a metabolism that varies oxygen demand and activity with temperature. This process is called oxyconforming, and in warmer conditions oxyconformers assume a sluggish, inactive lifestyle, which requires less oxygen. Many marine invertebrates adopt this metabolic strategy, but for more active animals, such as fish, it is not really an option— they are oxyregulators; they have a great need for oxygen at all temperatures. For the oxyregulators of the Early Triassic, this metabolic necessity was to prove their Achilles' heel.

Sun Yadong was the first to notice that something was amiss with fish in the Early Triassic despite all their evolutionary success. While picking conodonts for his temperature study, he noticed that there were no fish bones and teeth. This situation is highly unusual because small bits of fish bone typically co-occur with conodonts (which are also small bits of bone material). Searching the published literature on Early Triassic fish, we found that although they were abundant in high latitudes, very few were found in the tropics. This is even more surprising given that seafloor anoxia was prevalent at this time. Such conditions usually provide exceptionally good conditions for fish fossilization. We surmised that the tropics were simply too hot at this time for fish to survive. In contrast, at higher latitudes, where temperatures were cooler and everything was fine, fish were undergoing the exuberant diversification described above. Searching the literature further, we found that there was also a tropical gap for marine reptiles (which also diversified rapidly during the Early Triassic, but only in higher latitudes) and terrestrial vertebrates. Once again, this was not because of a lack of suitable sediments. On the contrary, lake sediments (which provide excellent conditions for terrestrial fossil preservation) were widespread at this time in central Europe and North China.

Paleontologists have spent many years looking for vertebrate fossils in these deposits but have turned up nothing.

Finally, our literature search has revealed that no amphibians are known from latitudes below 30° whereas they were highly successful at higher latitudes. This is a truly remarkable and fundamental contrast with the modern distributions of amphibians. Today the frogs and toads and their kin are diverse at equatorial latitudes, but their diversity rapidly declines outside the tropics. This has always been the way with amphibian distributions, but in the Early Triassic the opposite distribution was found: the amphibians had been driven to the poles. Once temperatures cooled in the Middle Triassic, fish, marine reptiles, and land animals, including amphibians, all returned to the tropical latitudes in great abundance.

High temperatures are also lethal for plants because as values climb above 35°C, photosynthesis becomes increasingly inefficient and photorespiration takes over as the dominant process. Modern observations on marine algae, such as those that live symbiotically in corals, show that this is a problem at even lower temperatures, around 30°C to 32°C. In contrast, one group of photosynthesizers, the cyanobacteria, can thrive at much higher temperatures; some have even been recorded photosynthesizing at an incredible 73°C. Therefore, the dominance of microbial (cyanobacterial) carbonates during the Early Triassic in the equatorial Tethyan Ocean could well be a response to high temperatures. Cyanobacteria may also have dominated the equatorial marine phytoplankton.

High temperatures may also explain the paucity of plants and the "coal gap" at this time. Plants probably struggled to thrive in excessive heat, but plants also decompose faster with increasing temperatures, thereby inhibiting the formation and preservation of peat and ultimately coal.

The Early Triassic global environment was in many ways a world turned upside down. The amphibians, which are typical of lower latitudes, were thriving only at the poles, whereas the tropics, normally the cradle of evolution and diversity, were an evolutionary desert: all the evolution was occurring at high latitudes.

How had the world got into such an appallingly hot state? In comparison, other rapid warming events—including those triggered by volcanism—have not lasted so long nor been nearly so intense. But in the Early Triassic, something seems to have been wrong with the normal feedback mechanisms that prevent a runaway greenhouse climate. Computer models have suggested that the spread of aridity within Pangea may have been a factor. Under humid conditions, rocks are weathered by rain laden with dissolved CO_2. The gas being in solution both draws down its atmospheric levels and increases the alkalinity of runoff into the oceans. Under more arid conditions this important process becomes less intense, and so the world remains warm. The evidence for aridity is a lack of plants, but another factor may be responsible for the absence of greenery.

Usually geologists reason that if there are no fossil plants, then the climate is likely to have been arid. But in the Early Triassic, there may have been another reason: the lack of plants may be an indicator of excessive heat and severely retarded photosynthesis. It may have been that the world was too hot for plants, especially in the lower latitudes of Pangea. Without plants there would also have been a lack of terrestrial organic-carbon burial (i.e., no peat swamps). The absence of coal formation is clear evidence for this state of affairs. Normally, organic-carbon burial on land is responsible for removing around half the CO_2 that goes in to the atmosphere, and the loss of this process may partly account for the ultra-greenhouse climate.

There may also be a threshold effect at work here to further exacerbate the Early Triassic hothouse. Once temperatures exceeded a critical value, the decay of organic matter may have been so fast that little survived to be buried in soils and marine sediments, thus greatly weakening the atmospheric CO_2 drawdown (remember that we keep the organic matter that we consume, i.e., food, in a fridge to prolong its freshness). It is noteworthy that the only locations where organic-rich sedimentation could be found in the Early Triassic were deep oceans, deep basins, and high-latitude shelf seas. These may have been the only places cool enough for significant organic-matter burial. With carbon burial in such a parlous state, it might be expected that any further CO_2 additions to the atmosphere would have exacerbated an already over-heated situation. This is exactly what seems to have happened 2 million years after the Permo-Triassic mass extinction.

THE SMITHIAN/SPATHIAN EXTINCTION

The Smithian and Spathian are the third and fourth stages of the Early Triassic respectively, and their boundary, around 2 million years after the Permo-Triassic boundary, marks what appears to have been another terrible crisis for life. For obvious reasons it has acquired the rather sinister-sounding names of the S/S crisis or S/S extinction. The victims were not especially numerous or diverse, because life, especially on the seabed, had barely begun to recover from the previous mass extinction—in other words, there was not a lot around to kill at this time. However, many of the groups that had thrived post mass extinction succumbed during the S/S event. The mollusk victims included the bellerophontids, an unusual group of snails that had thrived in the earliest Triassic, and several types of bivalve. The ammonoids once again underwent a drastic pruning of their evolutionary tree at the end of the Smithian although, as usual, they bounced back very quickly. The conodonts were also victims. They had done well at the end of the Permian, but they then suffered devastating losses at the S/S boundary, almost to the point of being wiped out; however, the latter fate would not occur until the end of the Triassic.

On land the S/S extinction has not been studied in detail, but it appears that *Lystrosaurus* may have died out at this time. Among plants, the clearest changes seem to have occurred during the Spathian Stage, immediately after the extinction. Thus, high-latitude areas saw the reestablishment of gymnosperm woodland—trees were back. The contemporaneous sections in South China similarly record a change from a low-diversity flora dominated by herbaceous lycopsids and mosses to a woodland flora dominated by conifers. However, no such change is seen in central European areas, where the tough, lycopsid shrub *Pleuromeia* continued to thrive, to the near-exclusion of all other plants, until the end of the Early Triassic.

The carbon and oxygen isotope changes around the S/S boundary show a series of changes that provide clues to the cause of the crisis. The $^{13}C/^{12}C$ ratio in limestones decreases significantly as the proportion of ^{12}C becomes greater. It may be recalled that a similar trend was also seen during the Permo-Triassic mass extinction, but this time, the swing to light carbon values was even greater. The oxygen isotope record from conodonts shows a parallel shift to lighter values and indicates that sea-surface temperatures, which were already hot, got even hotter, reaching a scalding 40°C in the latest Smithian Stage. These trends all sound like the Permo-Triassic boundary changes all over again, albeit somewhat magnified, and the same culprit—Siberian volcanism with CO_2 release—has generally been blamed once again. However, the detailed timing of the eruption history of the Siberian lavas has not yet been evaluated in detail, and so far the link between the S/S crisis and volcanism is best described as possible but unproven.[14]

Both the causation and the consequences of the S/S crisis are a little unclear and tricky to understand. Hot seawaters and widespread anoxia were prevalent throughout the Early Triassic, and the conodonts, among other groups, seemed to survive successfully while this was going on, so why did they succumb to these conditions at the end of the Smithian? Perhaps the S/S extinction is simply the peak intensification of these (very hot, very anoxic) conditions? Certainly temperatures of 40°C would be very difficult for many organisms to tolerate, but our data come from equatorial waters. Presumably values were a bit cooler at higher latitudes, and so this area should have provided a refuge from equatorial cooking, and yet it did not.

Thomas Galfetti and a team from Zurich have argued that an entire range of environmental stresses driven by volcanic CO_2 release into the atmosphere caused the S/S crisis. This is the "Murder on the Orient Express" explanation again, but they have generally focused on ocean acidification as the number-one cause. As with the list of Permo-Triassic extinction culprits, this is a highly controversial choice, but it is eminently testable because the effects of this stress vary with latitude. Warming and anoxia will be at their most severe in low latitudes, but in contrast, acidification should be a more effective killer at high latitudes. The reason is that cooler waters can dissolve more gas, such as carbon dioxide, and thus the decline in pH will be larger at higher latitudes. This scenario is illustrated by changes in the modern oceans over the past few decades as atmospheric CO_2 concentrations have increased, with the result that polar waters have shown significant decreases in pH whereas tropical waters have changed little. As it happens, our current state of knowledge regarding the S/S extinction makes it difficult to evaluate any latitudinal extinction bias, but it is noteworthy that one of the worst affected groups, the conodonts, do not have a carbonate skeleton and so should have been little affected by acidification.

So where does this leave us with regard to the S/S extinction? There is much work still to do and much to understand, but the spread of anoxic deposition in the oceans was so widespread that it must surely have played a role in limiting the available habitat for life on the seabed. The simultaneously increasing metabolic demand for oxygen driven by the increasing temperatures would only have made things worse for all life in the oceans. The nature of the crisis on land is unclear, but the apparent start of recovery in the aftermath of the S/S extinction suggests that conditions finally began to ameliorate after the acme of the super-greenhouse; the world was at last returning from the brink of a lethal hothouse.

SMOOTH SAILING IN THE LATER TRIASSIC

Within the 80-million-year time span covered in this book, the 45-million-year interval between the start of the Middle Triassic and the end of the period was probably the most successful for life. It saw the rise of a remarkable range of plants and animals, many of which are still with us today. In the oceans, seafloor life began its postponed recovery while fish and ammonoids continued their already successful radiations. Bivalves are some of the most abundant shellfish today, and their current success began with a major radiation in the Middle Triassic. This span witnessed the appearance of the first oysters, which still thrive, and the much less familiar megalodontids. Despite sounding like something that could join battle with the Transformers, the megalodontids were actually large, chunky bivalves that thrived in shallow equatorial waters. By the Late Triassic they were diverse and sufficiently abundant to start forming reefs alongside scleractinian corals—another group that appeared and diversified around the same time.

If you were to randomly pick a bivalve shell off a beach today, chances are it would belong to the venerids. These diverse and prolifically abundant clams include many of the types that we eat, such as cockles and quahogs. They were present in low numbers and low diversity before the Permo-Triassic extinction, but their long, slow rise began in the Middle Triassic. The success of the venerids seems related to their lifestyle because, unlike most pre-extinction bivalves, they live buried in sediment (an infaunal life position), rather than lying on the sediment surface (an epifaunal life position). Prior to the end-Permian extinction losses, the majority of marine invertebrates lived on the seabed, where they would have been visible to predators (and any time-travelling scuba divers). The venerids provide one of the many examples of successful infaunal groups from the Triassic. Today much seafloor life, including bivalves, gastropods, crustaceans, and echinoderms, live buried (and therefore hidden) beneath the seafloor. This fundamental change began after the mass extinction and is one of the clearest legacies of this 252-million-year-old crisis today. It may have been driven by the increasing predation intensity from newly evolved groups of shell-crushing fish and reptiles and the consequent need for seafloor life to hide within the sediment.

The Triassic seas were also host to a successful invasion by several terrestrial reptile groups. Of these, only the turtles remain today, but the other groups included the placodonts, which were seal-like reptiles with distinctive, flat teeth, capable of crushing shells; the nothosaurs, with long necks and pointy teeth, suggesting a fish-eating diet; and the thalattosaurs, with both a long neck and tail. The most successful and diverse of them all, however, were the ichthyosaurs, a reptilian equivalent to dolphins. These "fish lizards" first appeared in high latitudes in the Early Triassic (thereby avoiding the hot waters of the tropics), and even the earliest examples were fully adapted to marine conditions. Presumably ichthyosaurs evolved from a semiaquatic reptile that was something half way between a crocodile and a dolphin, but there is no fossil evidence for it.

The evolutionary transition from terrestrial to fully marine lizards must have been remarkably fast in all the marine reptile groups. Ichthyosaur evolution continued apace into the Middle Triassic, when numerous, specialized forms appeared. The study of this panoply of ichthyosaurian diversity is currently a highly active field, and many new forms are being described every year. In a 2013 paper, Nadia and Jörg Fröbisch from Humboldt University, Berlin, Germany, described an impressive, new genus from the Middle Triassic of Nevada. At nearly 9 meters in length, *Thalattoarchon* was much larger than a killer whale, and its mouthful of big stabbing teeth, up to 12 centimeters long, indicates that it ate large prey. The Fröbischs describe it as a "macrophagous apex predator," which sounds very impressive even in such technical jargon. Other Triassic ichthyosaurs included elongate serpentine forms, such as *Utatsaurus*, and much smaller, more dolphin-like forms with huge eyes, which surely were adapted to low light conditions in deep water. However, most impressive of all were giant, deep-bodied forms with long flippers. These include the 15-meter-long *Shonisaurus*, which can be seen in an excellent hillside museum in the Berlin-Ichthyosaur State Park in the Shoshone Mountains of Nevada. Even larger, *Shastasaurus*, the true giant of Triassic seas, approached 20 meters in length and had a highly elongate snout with very small teeth. It may have hunted giant squid, as the similar-sized sperm whales of today do. The ichthyosaurs' success was to last for 100 million years, but only at the start of their history in the Triassic did they show such a great range of forms and sizes.

The Middle Triassic recovery in the sea was also paralleled by evolutionary successes on land. Thus, well-studied animal communities from the Urals in Russia consisted of large herbivores (kannemeyerid dicynodonts, which had replaced the earlier dicynodont *Lystrosaurus*), small herbivores (cynodonts), and several groups of small carnivores, including the therocephalians and euparkerids. The last group resembled small, bipedal dinosaurs, which is not surprising because they are considered close, albeit primitive, relatives of this famous group. Erythrosuchids—big, beefy, quadrupedal predators—occupied the top of the terrestrial food web and were among the first of a whole series of impressive Triassic predatory archosaurs (a group that includes lizards and crocodiles today).

Evidence from plant communities reveals a similar burst of diversity increase at the start of the Middle Triassic, exemplified by the plant fossils collected from the Grès à Voltzia, a highly fossiliferous sandstone from northeastern France (which also yields some of the oldest fossil flies). Its plant fossils include conifers, horsetails, ferns, and ginkgos; the ubiquitous Early Triassic *Pleuromeia* was now gone. The ginkgos were to become common trees of the Triassic and Jurassic alongside the bennettitaleans mentioned above. However, unlike the latter, the ginkgos still survive—just. The maidenhair tree, *Ginkgo biloba*, is the sole survivor of this Pangean plant dynasty. Originally restricted to a few areas in China, it is now a common ornamental tree throughout the world, and its seeds are eaten and used in traditional medicines.

Many of the Middle Triassic plants were survivors of or had close relatives before the Permo-Triassic crisis, and so their reappearance around 10 million years after the extinction raises the question, where had they been? Some at least survived the Early Triassic in the cooler climes of Gondwana. However,

Plate 1. Bedding plane covered in Permian brachiopods (Spitsbergen, Norway), which were soon to be victims of the Capitanian mass extinction.

Plate 2. David Bond pointing his rifle at the point where Capitanian brachiopods go extinct in Spitsbergen, Norway.

Plate 3.　Simon Bottrell (University of Leeds) admiring some pillow basalt lavas of the Emeishan large igneous province, Yunnan, China.

Plate 4. The A team enjoying fieldwork in western Yunnan, China. Left to right: David Bond, Mike Widdowson, Sun Yadong, and Dougal Jerram.

Plate 5. The Permian-Triassic boundary seen in Anatolia, Turkey. Well-bedded fossiliferous limestones in the lower part of the cliff face give way to massive microbial limestones of Triassic age.

Plate 6. Sausserberget Mountain top in central Spitsbergen, Norway. The geologists walk on dark shales of the Triassic Period, which contrast with the underlying cliff-forming cherts of the Permian.

Plate 7. Early Triassic bivalves, some of the few fossils that are common at this time.

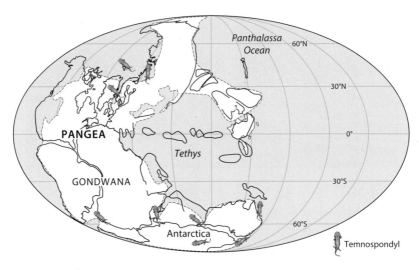

Plate 8. The Early Triassic world showing the distribution of temnospondyl amphibians, which were avoiding overly hot equatorial latitudes at this time.

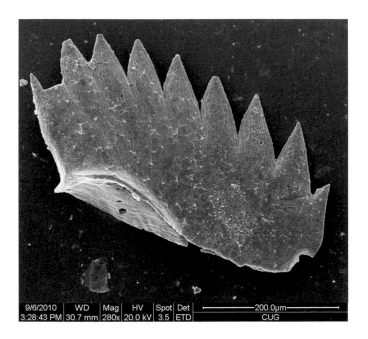

Plate 9. *Neospathodus*, a conodont from the earliest Triassic of South China, collected by Sun Yadong. Less than 0.5 millimeters in length, such fossils have proved very useful for dating rocks.

Plate 10.   A quarry face in a Late Triassic reef, Bavaria. The finger points to valves of megalodontid bivalves, and the white tubes are scleractinian corals. All species in this reef were wiped out by the end-Triassic mass extinction.

Plate 11.   A mountain of Triassic rock seen in the Dolomites of northern Italy. Sandstones of Carnian age form the lower cliff face, and the upper cliff is formed of Norian limestone.

Plate 12. Rob Newton examining beautifully laminated shales of the Early Jurassic Period, Somerset coast, England.

Plate 13. A block of sandstone from the Late Triassic Period, Somerset coast, England, with the inevitable hammer resting alongside. Earthquakes linked to the massive volcanic activity at this time probably caused the contortions of the layers in this rock.

Plate 14. Kettleness cliff, Yorkshire coast, England. The Early Jurassic (Toarcian) extinction occurs just beneath the top of the cliff, where soft-weathering shales give way to harder, black shales called the Jet Rock.

Plate 15. Bullet-shaped belemnites and an ammonite seen in Toarcian shales, Yorkshire, England. The belemnites are beloved of geochemists because they provide information about seawater temperatures.

Plate 16. Flood basalt landscape seen at Gásadalur in the Faroe Islands. These lavas erupted 60 million years ago and only had a modest effect on the climate, unlike the more devastating eruptions at the time of Pangea.

the main evolutionary success stories in southerly latitudes were shown by the corystosperms (seed ferns), of which *Dicroidium* was by far the most abundant, and the cycads and bennettitaleans.

An increase in diversity is one measure of recovery, and on this criterion alone the Middle Triassic was a good time for plants. However, plant abundance—known as the biomass—had still not recovered, as reflected in the continuing absence of coal deposits. Biomass-rich, peat-swamp communities only restarted in a fitful way, more than 12 million years after they had disappeared, toward the end of the Middle Triassic. The reason for this is not clear, but it has its counterpart in the failure of reefs to recover until the Late Triassic. For reefs, which are among the most complex of marine communities, their delayed recovery may reflect the considerable evolutionary time needed to evolve such complexity. Perhaps the same argument applies to luxuriant forest communities too?

Clearly, all was not plain sailing in the post–Early Triassic world but, nonetheless, by any measure it was a time of successful recovery. There is, however, one more remarkable and little-known climatic event that remains to be documented. It occurred in the Carnian Stage of the Late Triassic, around 20 million years after the Permo-Triassic mass extinction, and it had an extraordinary effect on life, from the smallest to largest of the world's organisms.

THE CARNIAN PLUVIAL EVENT

The term "pluvial" refers to rainfall, and so this event is named after an interval of heavy rain that occurred around 232 million years ago. It was short-lived by geological standards (perhaps a few tens of thousands of years) and punctuated a prolonged period of much more slowly changing climate.

Following the hothouse of the Early Triassic, the world gradually cooled down to more normal temperatures that were probably not dissimilar to those of today. As the Pangean supercontinent very gradually drifted northward, it saw the spread of humid equatorial areas. This expansion probably reflected the development of an optimal monsoonal climate as both arms of Pangea came to perfectly straddle the equator. The world sea level was also rising and thus caused some of the low-lying basins of Pangea to be flooded for the first time since the Late Permian, bringing more humidity into the continent's interior. Computer modeling studies of this interval suggests that the increased global rainfall in the later Triassic would have accelerated rock weathering and thereby raised the rate at which CO_2 was removed from the atmosphere. Therefore, in summary, a cooling trend in the later Triassic can be explained by the increased monsoonal conditions and Pangea's drift. But superimposed on the slow trends of cooling and increasing humidity, there was the short-lived and highly enigmatic Carnian Pluvial Event.

The bizarre and fascinating pluvial event was discovered and documented by two PhD research students at Birmingham University in the late 1980s. I had a ringside view of the birth of their idea because it took place in my office. At the time, Mike Simms and Alastair Ruffell were, like me, working under the supervision of Tony Hallam in a nice, spacious, high-ceilinged office at the top of the geology department. By rights, they should not have been the least bit interested in the Triassic climate. Mike was studying Jurassic fossils, and Alastair was supposed to be investigating Cretaceous sea-level changes; however, both have inquisitive minds and a talent for connecting apparently disparate observations. Mike had been considering the fate of Triassic crinoids—a group of echinoderms that had suffered badly during the Permo-Triassic extinction but were gradually recovering in the Triassic. He found that a lot of crinoids went extinct in the middle of the Carnian, an interval that could, up until that point, only have been a contender for the "dullest interval of geological time" award. He happened to chat to Alastair about his discovery of this crinoid crisis. Now as it happens, Alastair had spent part of his childhood in Somerset in southwestern England and knew his local geology, which consisted of Triassic outcrops. These had accumulated in the center of Pangea, and for the most part they record very arid deposition, mostly fine sediment that got trapped in ephemeral lakes in a desert. The Bonneville Salt Flats of Utah provide a nice modern analogue. The exception to this Triassic aridity in Somerset is the North Curry Sandstone, which was formed by rivers. Thus, there appeared to have been a strange, atypical phase of rainfall in the middle of Pangea during the Carnian. Amazingly, given these two incredibly disparate facts, Mike and Alastair were able to concoct a story that they published in the highly prestigious journal *Geology* in 1989.

The Simms-Ruffell hypothesis goes like this: during the middle Carnian, the world experienced a sudden, dramatic increase in rainfall. The resultant rivers shed vast amounts of sediment into the seas fringing Pangea, causing limestone formation to cease because all of the carbonate-producing organisms were swamped by sediment. Included were the poor crinoids, only just getting over their Permo-Triassic crisis. Of course these ideas left many questions unanswered such as, why did it happen? Mike and Alastair speculated that it may have been related to continental rifting and/or a major volcanic episode, but it was to be more than a decade before other scientists turned their attention to this climatic enigma and its causes.

The Carnian Pluvial Event is most spectacularly manifest in the mountains of northern Italy, where there is a great thickness of Triassic limestones to be seen. They form particularly impressive scenery in the Dolomite Mountains, which are famous for their winter skiing and also the 1990s Sylvester Stallone blockbuster *Cliffhanger*. The only interruption to the long history of limestone deposition occurred in the middle of the Carnian, when rivers washed sand into the seas to form sandstone beds that can be seen (if you know where to look) as distinctive ledges in some of the *Cliffhanger* footage. The Dolomites sandstone, like the Somerset sandstone, is recording rainfall only in western Europe, and it could easily be argued that it records climatic changes in this region—perhaps caused by an intensification of the monsoonal climate. However, as Simms and Ruffell suspected in 1989 and subsequent study has shown, the Carnian Pluvial Event was much more widespread. Recent investigation of clay minerals in the deep-ocean sedimentary rocks of Japan has found a bed of clay in the Middle Carnian, and the mineral is of a type that forms in humid climates. The Japanese clays settled in the Panthalassa Ocean, having been blown far out to sea, some thousands of kilometers from Pangea, and their presence shows that the pluvial event had a very far reach indeed.

The Carnian Pluvial Event is a fascinating episode of the Triassic, and I wish I had played a part in its original discovery, but my excuse is I was too busy working on my Jurassic-focused PhD. At least I have been peripherally involved more recently in a project led by Nereo Preto at the University of Padua in northern Italy. Together with his research student Jacopo Dal Corso, Nereo has shown that the event has many of the hallmarks of other Pangean mass extinction episodes. Most notably, there was a major (but short-lived) change in carbon isotopes to lighter values. As we have already seen in chapter 3, a similar carbon-isotope change during the Permo-Triassic mass extinction is explained by the emission of light carbon from the Siberian eruptions with consequent rapid warming effects. The same cause and effect has also been suggested for the S/S crisis. We do not know if there was also warming at the time of the Carnian event, I suspect there was, but several teams are currently looking for the evidence.

There is a possible volcanic culprit for the Carnian climatic changes. The Wrangelia Province is a vast region of flood basalts that form a large part of the coastal mountain ranges on the western seaboard of Canada. The province originally formed as a large oceanic plateau within the Panthalassa Ocean. Such plateaus are the oceanic equivalent of the continental flood basalts, like the Siberian and Emeishan Traps, but because they sit on oceanic crust, they are inexorably carried to subduction zones. Once there they get partially dragged down into the mantle, but substantial portions get stuck on the overriding continental edges. Current best estimates for the age of the Wrangelian basalts suggest that they were erupted rapidly around the middle of the Carnian.

So the Carnian Pluvial Event has many of the ingredients of a true extinction crisis: massive volcanism, a major change in the carbon cycle, and climate change. There are also extinctions. As well as the crinoid losses first noted by Mike Simms, several other marine groups—notably bryozoans, ammonoids, and conodonts—all suffered in the mid-Carnian alongside several marine reptile groups, including the thalattosaurs. On land several major groups disappeared: the rhynchosaurs, along with those hardy survivors of the Permo-Triassic extinction, the dicynodonts; although the precise timing of their demise is a little uncertain, a few may have survived into the Jurassic. Indeed, there has been a prolonged debate about the extinction losses among land animals in the later part of the Triassic. Thus, as long ago as 1983, Mike Benton noted that many land animals disappeared at some time in the Carnian. The losses included the chiniquodontids, a group of therapsids that looked a little like giant rats with oversized heads, and several groups of large-bodied amphibians. The loss of the latter animals is rather curious because wet habitats became widespread during the Carnian pluvial event, a change that should have been favorable to such water-loving animals. Such was the scale of terrestrial extinction that many vertebrate paleontologists regard it as the equal of or even greater than the end-Triassic mass extinction, which will be discussed in chapter 5.

The Carnian extinctions are quite impressive, but the most profound and long-term effect of the pluvial episode is the remarkable impetus it gave to evolution: many major new groups appeared at this time. The first primitive dinosaur-like animals were on the scene before the Carnian, but they were small and essentially trivial components of terrestrial communities. All of that changed during the later Carnian, when dinosaurs underwent a major radiation and became a much more diverse and significant component of the fauna. The pterosaurs, their flying reptilian cousins, also appeared around the same time, as did the first mammals. *Adelobasileus*, the earliest representative of our own group, dates from the later Carnian, as does the first probable flowering plants—*Sanmiguelia* from the Southwest of the United States. Adding yet more diversity to the terrestrial landscape, the later Carnian also saw the appearance of diverse types of reptiles, the crurotarsan archosaurs, which would go on to prominence in the final 30 million years of the Triassic. They included a broad diversity of reptilian types ranging from the fish-eating phytosaurs (big, chunky, crocodile-like reptiles) to the armored herbivorous aetosaurs, with curiously small heads and large shoulder spikes. Scariest of all were the giant carnivorous rauisuchians. Up to 6 meters in length, they look like long-legged crocodiles on steroids. With their legs beneath their bodies, they were undoubtedly capable of galloping at considerable speed. Thus, the post-Carnian climatic episode was witness to an impressive diversification among animals, with the dinosaurs and crurotarsans vying for "top dog" status. This competition was to persist for the final 30 million years of the Triassic. With the benefit of hindsight we know who the winners of this contest were, but at the time, the crurotarsans were often the most abundant and diverse animals to be found in terrestrial communities.

Probably the most fundamental change to occur in the later Triassic was among the plankton. Immediately after the Carnian Pluvial Event, a planktonic group of green algae started secreting a tiny calcareous shell. This was the first time in the history of the oceans that photosynthetic plankton had started to make their own little skeletons of calcite. The result was that tiny carbonate particles started to rain down to the seabed and form calcareous muds. Such carbonate rain is a fundamental part of the ocean carbon cycle today, but what triggered its appearance in the Carnian? One possibility is that all the enhanced rainfall caused a major inflow of alkalinity to the oceans (due to the weathering of silicate rocks) saturating the oceans with calcium carbonate and making it possible for plankton to start calcifying. An advantage for algae is that for each molecule of carbonate precipitated, a molecule of carbon dioxide becomes available for photosynthesis. This major new innovation at the base of the marine food chain proved to be a false dawn because the calcareous rain soon stopped. Thus the Carnian deep-ocean carbonates are only a thin bed sandwiched between the normal radiolarian cherts on Panthalassa's floor. It was to be another 100 million years before such ocean floor sediments were to form again.

After the Carnian false start, calcareous plankton reappeared tens of millions years later, toward the end of the Triassic. This time the carbonate was produced by a new group of plankton called coccolithophorids, and they were one of several new planktonic groups that appeared at this time. They also include the dinoflagellates and various groups of ultratiny plankton called nannoplankton; all are loosely called red algae because they have chlorophyll c in their chloroplasts. In contrast, the previously dominant green algae use chlorophyll b. The green/red split probably occurred more than a billion years ago, but for the first 770 million years, green algae dominated ocean plankton. This changeover to red algae has been called one of the largest changes ever seen in the oceans; however, it was a gradual one, and not until the late Jurassic, 60 million years after the Triassic, could the transition be considered complete. By this time the coccolithophorids were widespread in the open ocean, whereas in the Triassic they had been restricted to the waters of warm shelf seas.

Today the coccolithophorids are key components of the carbon cycle. Their carbonate skeletons are responsible for removal of much atmospheric CO_2, while the presence of a vast blanket of coccolithophorid mud on the deep seabed provides a fantastic buffer against changes in ocean acidification. We will return to the importance of deep-water carbonate sediments in chapter 7, but in the Late Triassic, all this lay in the future. Like the dinosaurs, pterosaurs, mammals, and angiosperms on land, the coccolithophorids were the harbingers of a future world.

CHAPTER 5

TRIASSIC DOWNFALL

The final few million years of the Triassic saw the landmasses of South and North China gradually collide into one another. This was followed by their joint collision, together with other bits and pieces of Southeast Asia, into the northern coast of Pangea. The result was a Pangean supercontinent at its zenith; all the pieces of the jigsaw were together (fig. 5.1). It would now have been possible to walk on every piece of continental crust without once having to cross an ocean. No need for boats.

While continental drift was making world geography very simple, life was thriving, having long forgotten the earlier extinctions of Pangea's history. Following the enigmatic Carnian climatic episode, terrestrial communities were diverse and consisted of large crurotarsans and, to a lesser extent, dinosaurs, plus rarer groups such as the pterosaurs and mammals. Forests were back on the scene and well established, and plants were abundant enough to form coal deposits. Insects were starting to look very modern, with flies and beetles evolving fast alongside more exotic groups like the titanopterids— giant, overgrown grasshoppers with wingspans in excess of 20 centimeters. In the seas coral reefs had appeared in the warm waters of western Tethys, while plankton was undergoing a fundamental change in composition as red replaced green in the great algal changeover. However, this is a book about great

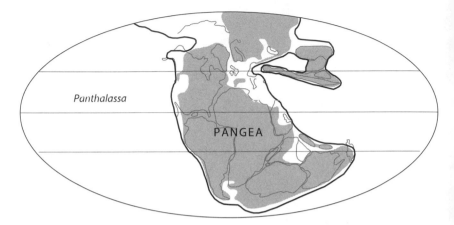

Figure 5.1. World map at the time of the Triassic-Jurassic boundary, 200 million years ago, showing the Pangean supercontinent at its peak size.

disasters, not successes, and once again the restless nature of plate tectonics was to be the undoing of evolution's hard work. Another gigantic phase of world-ending volcanism was about to strike, this time centered in western Pangea.

Andrea Marzoli of the University of Padua, in Italy, is one of the foremost volcanologists to investigate a LIP that he has termed the Central Atlantic Magmatic Province, or CAMP for short. This name is somewhat misleading (or puzzling) because the Atlantic did not exist when CAMP started erupting. However, Andrea chose this epithet because the site of eruption eventually became the central part of the Atlantic as the fragments of Pangea drifted apart over tens of millions of years. A consequence of the separation and drift is the fragmentation of formerly united lava fields. For a long time this meant that the unity of the original province was overlooked, but recent study of the dispersed fragments has allowed geologists to piece together the extent of a formerly huge LIP—its lavas may have exceeded 5 million cubic kilometers in volume.

Two hundred million years later (i.e., today), the effects of erosion and burial have left only remnants of CAMP's igneous rocks, although some, such as the Palisades Sill of New Jersey and New York, are still impressive. Where the basalts have been eroded, the remaining igneous plumbing system bears testimony to the magnitude of the event: dikes—vertical cracks filled with cooled magma—in Amazonia, for example, are 200 meters wide and 300 kilometers long. A single dike in the Iberian Peninsula has been traced for more than 500 kilometers.

With the identification of the huge scale of CAMP volcanism in the late 1990s came the realization that its being 200 million years old placed it in the frame for the end-Triassic mass extinctions. Although relatively little studied at the time, the crisis has long been recognized as one of the big five, and the identification of such an enormous smoking gun was to spark an intense research effort that continues to this day. Nowadays the end-Triassic catastrophe is one of the most actively investigated of all mass extinctions. The link with LIP volcanism inevitably causes many parallels to be drawn with the Permo-Triassic mass extinction/Siberian Traps nexus, and there are indeed many close similarities. But before racing on to consider the question of how it happened, we must first look at what died and when.

THE VICTIMS

Dinosaurs, the iconic group of paleontology, ruled the world for the 135 million years of the Jurassic and Cretaceous Periods, but as we have seen, they had been around for tens of millions of years prior to this, which raises the question, why did it take them so long to become dominant? The answer seems to be that it took a mass extinction to remove their chief competitors, the crurotarsan groups—such as the aetosaurs, the rauisuchians, and the phytosaurs. These diverse animals were present right up to the end of the Triassic, but none are known to have survived into the Jurassic. The survivors included true crocodiles (the only surviving legacy of the crurotarsan dynasty), pterosaurs, and mammals, but it was the dinosaurs that were able to respond with alacrity to the emptied environments and went on to rapidly fill the earliest Jurassic landscape.

The idea that the crurotarsan/dinosaur transition was caused by a mass extinction was a highly controversial notion when initially suggested more than thirty years ago. Many had thought the Late Triassic had witnessed a prolonged competitive struggle with the dinosaurs finally coming out on top. Subsequent work has now shown that the crurotarsans were in fact doing well until near the end of the Triassic—it needed a catastrophe to remove them from the stage. The dinosaurs therefore owed their success to a combination of luck and an ability to survive the harsh conditions of the extinction interval. However, even if the nature of the transition is now well accepted, the precise timing is still debated. Here we need a digression on the divisions of Late Triassic time. This interval is subdivided, from oldest to youngest, into the Carnian, Norian, and Rhaetian Stages (see fig. 1.1). Some have argued (notably Spencer Lucas of the New Mexico Museum of Natural History) that the main terrestrial animal extinctions occurred at the end of the Norian Stage, whereas others suggest it happened later, toward the end of the Rhaetian Stage. It would be very useful to resolve this debate because the marine extinctions (described below) happened in the late Rhaetian and so may (or may not) coincide with the terrestrial crisis. Our understanding of Late Triassic time is not helped by the great uncertainty regarding the duration of the Rhaetian Stage. Estimates range from a mere million years to as much as 10 million years. The larger estimates make it possible that the terrestrial extinctions may have substantially predated the marine extinctions if the latter occurred at the end of the Norian. This would mean that the two events had nothing to do with each other but were in fact totally separate extinctions—an untidy conclusion. On the other hand, a very short duration for the Rhaetian Stage would mean that the marine and terrestrial crises occurred about the same time, so we could be talking about a single mass extinction.

Some remarkable and pertinent evidence to help resolve the debate comes from southwestern England and Wales, and it is found in Carboniferous limestones. This may sound strange, because Carboniferous rocks formed 150 million years before the Late Triassic. However, by Rhaetian times the limestones in this region had been uplifted to form small islands up to 16 kilometers in extent, which were surrounded by a warm shallow sea. Importantly, the limestone landscape was riddled with caves and fissures. As a result, animals walking around the islands often fell down holes, got trapped, and thereby became fossils. This fissure fauna, as paleontologists have called it, contains a remarkable diversity of animals dominated by small lizards called sphenodontians and other reptile groups. Unfortunately though, it does not provide a complete census of latest Triassic life but rather a subset: a specialized island-dwelling community. It notably lacks the larger crurotarsan animals, probably for the simple reason that it is hard for a big animal to fall down a small hole. However, there are rare finds of somewhat bigger beasts, including aetosaurs, phytosaurs, and dinosaurs, some of which approached two meters in length, so some of the crevices must have been quite large. Mammals, on the other hand, are absent despite their small size, probably because they are not very good at dispersing to islands even today. The high metabolic rate of mammals requires that they eat and drink every day, whereas the process of island colonization often requires that animals be clinging to driftwood for many days or even weeks. This is something that reptiles but not mammals can survive.

Despite the odd composition of the Rhaetian fissure faunas, it significantly contains many animals that failed to survive into the Jurassic. This suggests they are the record of a pre-mass-extinction island community and so supports the idea of a latest Rhaetian age for the terrestrial extinctions. We also have evidence from plant remains that the Late Triassic mass extinction happened in the latest Rhaetian and not earlier.

Sporomorphs (fossil spore and pollen grains) provide much better detail on the age of the end-Triassic extinction because they fossilize easily and are found in many types of sedimentary rocks. The best sporomorph data has come from studies of Triassic-Jurassic boundary sections in northwestern Europe and Greenland. These reveal a late Rhaetian crisis in which there were many interesting changes. Conifers, cycads, and bennettitaleans dominated the Late Triassic forests, but this situation abruptly changed in the later part of the Rhaetian, when fern spores become prolific. This "fern spike" was short-lived though, and conifer-dominated forests reappeared before the end of the Triassic. However, things had changed; the post-fern-spike forests were noticeably less diverse and their composition was different. The main conifers were now the cheirolepidiaceans, an extinct group that resembled modern cypress trees and were thought to prefer warmer and more arid conditions. So, like the plant crisis during the Permo-Triassic transition, the European and Greenland forests disappeared in the Late Triassic, but unlike the Permo-Triassic event, it was a brief phenomenon with plenty of changes and less extinction. The Triassic-Jurassic crisis may also have been of more regional extent.

The sporomorph story from North America is similar to that in northwestern Europe and Greenland, which is perhaps not that surprising given that they were all in the same region of tropical western Pangea in the Late Triassic. Significantly though, plants from more distant locations show much less dramatic changes. Thus, the Junggar Basin of northwestern China lay in northeastern Pangea, and it contains a thick and monotonous pile of fine-grained lake sediments that formed during the Late Triassic and Early Jurassic. Neither the sediments nor the fossil spores and pollen change much at boundary (although a short-lived fern spike may be present), suggesting that there were no major, long-term climatic changes in this area. Similar muted changes in Australia point to the same story in southeastern Pangea. The dramatic change in plant communities in central Pangea may relate to their proximity to the CAMP volcanism and its regionally damaging effects. More distant locations appear to have weathered the crisis somewhat better.

Let's stay for a moment with the ever-so-useful sporomorph studies. The extraction of such tiny organic fossils from sedimentary rocks also releases other bits of organic detritus, scattered among which are charcoal fragments—the product of past fires. As we look at charcoal amounts, it appears that there was a marked increase in the latest Triassic that persisted for a million or so years into the Jurassic. The evidence for increased burning is also supported by the presence of organic molecules called polycyclic aromatic hydrocarbons (PAHs), which are produced by combustion of vegetation. So, what can we infer from the charcoal/PAHs evidence? Once again the data come from central Pangea, where we know there were substantial vegetation changes. The more frequent wildfires likely reflect increasing aridity and more flammable plants in the changed forests. Some researchers have suggested that it was also a cause of the terrestrial mass extinction. Death by fire certainly sounds like a suitably hellish way of causing extinction, but it is a highly unlikely proposition.

It certainly gets no support from the study of modern terrestrial ecosystems. Many habitats actually benefit from fires because they cause openings in forest canopies and increase plant diversity.

Plant fossils, especially of leaves, can also provide clues to atmospheric changes. Leaves have tiny holes in their surface called stomata, which are used for gas exchange—during photosynthesis they let carbon dioxide in and oxygen out. This arrangement is very useful, but the downside of stomata is that they also let water vapor escape and cause desiccation of the plant. The density of stomata on the leaves of plants is therefore a careful balance between having enough for gas exchange but not so many that the leaves dry out too easily. Now when atmospheric carbon dioxide levels are very high, plants can get by with few stomata, but when the levels of the gas fall, then they need more. The stomatal abundance on leaves is thus a monitor of atmospheric carbon dioxide concentrations—a very useful relationship for those who study past climates. Care has to be taken though; for example, if the fossil-leaf record shows a doubling of stomatal density, it is safe to infer that carbon dioxide levels increased, but it does not mean they have doubled; some plants are just more responsive than others. So when dealing with changes in stomatal density, it is important that the same type of plant is being studied, something that is not always possible. Despite such caveats, any method for evaluating past atmospheric gas concentrations is highly desirable, and many studies have looked at leaf stomata across the Triassic-Jurassic transition. Jenny McElwain of University College Dublin has led much of this work, and she has shown that carbon dioxide levels rose steeply in the latest Triassic to reach a peak of about 2500 ppm.[15] This high point persisted for around 300,000 years before declining in the early Jurassic. For comparison, preindustrial atmospheric CO_2 levels were 280 ppm, and they are now around 400 ppm. It is safe to assume that it got very warm at the end of the Triassic, much warmer than today.

The end-Triassic seas and oceans also saw substantial losses among a whole range of environments and organisms, from Tethyan reefs to the tiny, planktonic radiolarians that drifted in the surface waters of Panthalassa. The most detailed extinction stories have again come from Europe, particularly the coastal cliffs around Britain. These accessible outcrops record a rather unusual depositional setting. A few million years before the end of the Triassic, the sea level rose, causing a marine embayment to develop over a large part of northwestern Europe (leaving islands of limestone with potholes, which were a danger for the resident animals). This Rhaetian Sea invaded a vast and very flat desert plain, and consequently the sedimentary layers that accumulated were of remarkable persistence. I have studied cliffs of Rhaetian rocks on the Devon coast, in southern England, and around Larne, in Northern Ireland, and been able to trace exactly the same series of beds even though the two locations are 600 kilometers apart. In addition to this remarkable uniformity, this vast inland sea was also unusually saline, no doubt because it was surrounded by the deserts of Pangea and subject to intense evaporation. The hypersalinity is manifest by an impoverished fossil content. Animals that were accustomed to normal marine salinity, such as ammonoids, brachiopods, corals, bryozoans, and echinoderms, are mostly absent, whereas groups able to tolerate high (and fluctuating) salinities, such as bivalves and ostracods, are present in abundance. It is the bivalves that tell us most about the extinction in this end-Triassic sea.

The initial sediments to accumulate in the Rhaetian Sea consisted of black shales belonging to the Westbury Formation. These are overlain by the white limestones of the Lilstock Formation, which in turn are overlain by the Blue Lias, a formation of limestone and shale beds that are famous for

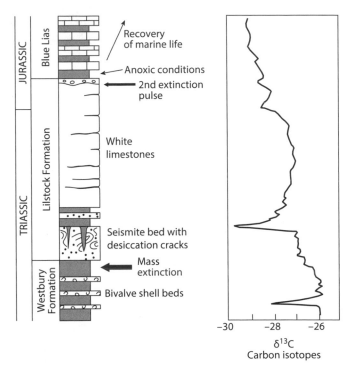

Figure 5.2. Triassic-Jurassic boundary rocks in southwestern England showing the levels of extinction (*left*) and the carbon isotope values (*right*) recorded in organic matter.

their Jurassic ammonites. The black, white, and blue stratigraphy provides information on a fascinating series of events (fig. 5.2). The Westbury Formation yields abundant and fairly diverse bivalves, but the majority of species (15 of 25) disappear at the base of the Lilstock Formation, thus marking the level of the mass extinction. About a meter above this level, the beds are intensely contorted in a way that is diagnostic of earthquake shaking. Geologists call such beds seismites, and just above them there is evidence for uplift and emergence in the form of deep desiccation cracks. The next limestone beds look innocuous enough in the field, but a study of the organic carbon isotopes at this level reveals that they are very light; that is, they are enriched in carbon-12 and depleted in carbon-13 compared to the beds on either side. This negative carbon isotope spike is a feature that we have met several times already during Pangea's earlier mass extinctions, especially during the Permo-Triassic crisis. This is a key level in the succession because it also corresponds to a change in the composition of spore and pollen fossils and is where the fern spike seen elsewhere in Europe and North America developed.

Higher in the Lilstock Formation, a few new bivalve species appear, indicating the start of recovery, but the real diversity bounce back was not until the beginning of the Jurassic. Prior to this, at the base of the Blue Lias a series of black shales and limestones developed, revealing that seafloor oxygenation was exceptionally poor at this time. There was clearly a lot going on in the late Rhaetian, but perhaps the most surprising aspect is that the first thing to happen—the bivalve mass extinction—predates everything else. By extending our view of the end-Triassic marine extinction to other regions we find that, like the Permo-Triassic mass extinction, the timing of losses varied around the world, and it may be more appropriate to talk of a two-phased crisis once again.

In southern Europe the Triassic-Jurassic boundary is found in shallow marine limestones that form lovely Alpine mountains in western Austria and southernmost Germany. The most profound change here is the extermination of diverse reefs constructed by scleractinian corals. The bivalves also had a tough time here too. Several groups of giant bivalves had evolved in the Late Triassic—the megalodontids, dicerocarditids, and wallowaconchids—and none of them survived into the Jurassic.

The Alpine limestones formed in the warm, shallow, equatorial seas at the western end of Tethys. If we follow the southern coastline of this ocean, we move gradually southward into temperate latitudes, where interbeds of limestone and terrigenous sediments (sandstones and shales) accumulated in shallow seas. Today some of these sedimentary rocks are found at great altitude in the Tibetan Himalayas, where they can be reached with difficulty. Tony Hallam and I visited one of the most accessible locations, close to the Tibetan border around fifteen years ago. I say "accessible" because the rocks were adjacent to the main Lhasa-Kathmandu highway, but this is a subjective term when used to describe the road infrastructure of 1990s Tibet. Tony has recounted some of our escapades on this trip in his book *Catastrophes and Lesser Calamities*. Other than the discomfort of spending two weeks sharing the back seat of a Mitsubishi Pajaro with three other geologists, I well remember taking a supposed shortcut over a mountain on a narrow, gravelly track that barely clung to the mountainside. When not spending time on death-defying journeys, our accommodation was in a Chinese army barracks, from which we had a splendid view of Mount Everest, but sadly we had no toilet facilities other than a narrow trench next to the parade ground. Overall the Tibetan visit was an ordeal, but it was worth it because it revealed a Triassic-Jurassic mass extinction story that is different from that of the Rhaetian Sea.

The Tibetan sediments formed in open marine conditions in which there was a high diversity of marine invertebrates. Usefully, they included the ammonite *Psiloceras*, whose first appearance defines the start of the Jurassic. Remarkably, most of the diverse Triassic bivalves survive for a short distance above this level, indicating that the end-Triassic extinction in this region actually happened in the earliest Jurassic. Detailed comparison with the European record is problematic, but it is possible that this later extinction in Tibet corresponds to a weak extinction pulse in Britain. Thus, a few bivalve species are lost at the base of the Blue Lias at the point of onset of black shale deposition (fig. 5.2). If so, this makes the chronology of the end-Triassic mass extinction incredibly similar to that of the Permo-Triassic—a double-phase extinction separated by a few hundred thousand years with the strength of each pulse variable from region to region.

Moving our attention to the open ocean, we find that the end-Triassic crisis was also manifest here. The radiolarians, which we last encountered struggling across the Permo-Triassic boundary, had rediversified and produced many new forms, including amusing hat-shaped species (hat-rads, if you like), and strange, fancy forms with twisted spines. All these types disappeared at the end of the Triassic to be replaced by radiolarians with much simpler, spherical morphologies in the Early Jurassic. Hat-rads lived in the surface waters and their demise indicates the severe nature of the stresses in the uppermost water column.

The radiolarian extinction had a brief but noticeable effect on ocean-floor sedimentation, which is now preserved in Japan. Red siliceous cherts made of radiolarian skeletons accumulated on ocean floors throughout the Triassic and Jurassic, but at the Triassic-Jurassic boundary this deposition was interrupted by a red mudstone that marks the temporary cessation of the rain of radiolarians to the ocean floor. If one looks down a microscope, the mudstone can be seen to contain fragments of volcanic glass and also pyrite—both components unique to this level. The glass suggests some far-distant volcanism (it is, of course, tempting to say this is a manifestation of the CAMP eruptions even though these were happening thousands of kilometers away from the middle of the Panthalassa Ocean). The pyrite suggests a brief pulse of ocean anoxia. Remarkably, it is also possible to infer climatic changes at this time even though the red mudstones were forming several kilometers below the ocean surface and thousands of kilometers from land. The evidence comes from subtle changes in the isotopic ratio of the trace metal osmium, which indicate a major but brief increase in runoff into the oceans at the end of the Triassic.

Other extinctions in the water column include the loss of the conodonts (a sad loss for paleontologists and geochemists, who love to analyze these fossils) and most of the ammonoids too. However, both of these groups appear to have been in decline throughout the Rhaetian Stage, suggesting that their final disappearance in the late Rhaetian was a coup de grâce for already impoverished groups. The real puzzle of the conodont and ammonoid decline is therefore, what was suppressing the usual evolutionary ebullience of these groups during the Rhaetian? Perhaps the Triassic-Jurassic extinction should be considered a protracted crisis? We have once again returned to the problem of not knowing the duration of the Rhaetian. If this stage only lasted a million years, then the entire extinction episode is brief, geologically speaking, but a 10-million-year Rhaetian implies an exceptionally long period of environmental stress. There is much work still to do to resolve these problems.

Ichthyosaur diversity also crashed in the later Triassic, although the precise timing is poorly resolved due to the rarity of fossils of these marine reptiles. Taking the fossil record at face value, it seems that many disappeared in the Norian Stage followed by the appearance of new types, such as *Ichthyosaurus*, in the Rhaetian. This hints that the Late Triassic crisis of water-column dwellers, such as the ichthyosaurs and ammonoids, was out of kilter with the other marine life. Overall though, for most groups the extinction was a late Rhaetian affair, but the larger water-column dwellers appear to have already suffered a (separate and earlier?) crisis by this time. In some regard, the end-Triassic mass extinction has parallels with the Permo-Triassic crisis, when nektonic predators were among the relatively few groups to be unaffected.

WHEN AND HOW DID IT HAPPEN?

When attempting to understand the end-Triassic mass extinction, we are faced with explaining many of the same attributes of earlier mass extinctions. This was an environmental disaster bad enough to kill vast amounts of life everywhere: the dominant animals on land were wiped out, forests disappeared, and the tiniest plankton in the oceans suffered greatly, and yet intriguingly it afflicted other groups less harshly. We also have the same potential culprit—a series of giant flood basalt flows. Can we therefore cut and paste the same extinction cascade (fig. 3.4) from the older to this younger event? The answer is yes, sort of.

First, how does the timing of the eruptions compare with the timing of the extinctions? Much effort has been expended on this question, and the answer depends on how the volcanism is dated. One approach is to directly date the lavas using tiny zircon crystals because they contain significant amounts of uranium that allow the use of uranium-lead dating methods. This approach reveals that the initial eruptions began in Morocco and rapidly expanded (probably within less than 20,000 years) to cover the Newark Basin of the eastern United States. This huge burst of eruptions happened at the same time as the fern spike; we could not ask for better evidence of synchrony between volcanism and extinction. However, it is a bit more complicated than that. There is other indirect evidence to suggest that eruptions may have begun earlier. Large-scale volcanism affects the trace-metal content of the oceans. Oceanic osmium is particularly susceptible to volcanic influence because the isotopic ratio of osmium atoms in the mantle is different from that in the crust. By interpreting a change in the osmium content of seawater (recorded in marine Tethyan limestones), Andrea Marzoli has suggested that CAMP volcanism began at the start of the Rhaetian Stage. A similar change is seen in the contemporary deep-sea cherts of Japan, giving further credence to Marzoli's notion. This would make the Moroccan lavas not the oldest of the CAMP but just the oldest that have been sampled so far. Some fragments of the LIP, such as little-known remnants of lava fields in the Amazon region of Brazil, could be older.

We therefore have two possibilities regarding the role of volcanism in the latest Triassic extinctions. Initial eruptions may have begun at the start of the Rhaetian (when pelagic groups such as ichthyosaurs and ammonoids began to suffer and maybe the crurotarsans were wiped out on land) followed in the late Rhaetian by a climax of eruptions and a major extinction on land and sea. Alternatively, eruptions may have begun in the late Rhaetian, and the earlier extinctions are either just misdated (e.g., the crurotarsans) or unrelated to volcanism (e.g., the ichthyosaurs and ammonoids). At present we cannot decide between these scenarios, although volcanism is accorded the main killers' role in both.

As for the case of the Permo-Triassic mass extinction, the principal candidates for end-Triassic kill mechanisms are the gas effusions from basalt flows. The volumes of carbon dioxide release were probably impressive, as they were with the Siberian Traps, but it is unclear if these alone were sufficient to cause devastating warming or whether additional greenhouse gases were needed. The negative carbon isotope spike, noted above, could be a sign of methane release from gas hydrates (as has been argued for the same trend seen earlier). Thus, a warming pulse triggered by volcanic CO_2 may have caused the release of methane, an effective greenhouse gas, and thereby exacerbated the temperature increase—a perfect positive feedback. Further stimulus may have come from the release of carbon dioxide from sediment baking beneath the

CAMP lava pile (Svensen's hypothesis again; see p. 34). As discussed above, there is certainly ample evidence from leaf stomata for this warming. Changes in vegetation recorded by sporomorphs also support a transition to warmer conditions.

Estimates of past temperatures can also come from fossil soil horizons,[16] and studies from the Newark and Hartford Basins have provided some intriguing resolutions to the Triassic-Jurassic boundary warm phase. Rather than being a single warm interval, it has been resolved into a series of rapid warming-gradual cooling oscillations, each lasting a few hundred thousand years. What is going on? The answer probably lies in the nature of lava flows and their tendency to weather. The latter is immediately apparent to anyone who visits modern lava fields, such as those in Hawaii, where even flows that erupted only a few hundred years ago have soil developing on their upper surface. The weathering of fresh volcanic rock consumes a lot of atmospheric carbon dioxide (which ends up as bicarbonate washed into the ocean), and so eruptions are in a sense self-limiting: they put carbon dioxide into the atmosphere, but they also take it out again as the lava weathers. There is a lag, though, because the chemical breakdown of lava takes tens to hundreds of thousands of years. Hence, the warming-cooling oscillations at the Triassic-Jurassic boundary probably reflect episodes of eruption followed by quiescence and weathering. Such cycles only occur if the lava flows are erupted on land, as happened with the CAMP basalts, and in distinct pulses. There is no such feedback if lava flows follow each other quickly, because they become protected from weathering by the next lava blanket, or they erupt beneath the sea, where they get buried in sediment. This last scenario appears to have been the case with the Siberian eruptions, which occurred at sea level, and may partly explain the prolonged, continuous, and catastrophic warming trend at this earlier time.

Was warming in any way directly responsible for the extinctions? Jenny McElwain has certainly proposed this, but among the majority of Triassic-Jurassic cognoscenti, the favored cause of marine death is ocean acidification. Anoxia is also accorded a role in some extinction models although it is usually accorded secondary status in kill lists after acidification. This double whammy is also popular for the Permo-Triassic mass extinctions, but as we saw earlier, the preferred order is reversed—anoxia is clearly implicated whereas acidification is less certain. Michael Hautmann (University of Zurich) was the first to propose acidification as the end-Triassic killer. He noted that in many Triassic-Jurassic boundary records from around the world, limestone beds temporarily disappeared and were replaced by shales, and he interpreted this change to have been caused by intense ocean acidity. This transition is not seen everywhere though. In Britain the extinction occurs at the transition from black shales to the white limestones of the Langport Member. Perhaps the unusual chemistry of the isolated Rhaetian Sea in this region ensured that it was somehow protected from changes in the ocean pH?

The extinctions among bivalves have also been used to support an acidification scenario. Bivalves construct their shells using either aragonite or calcite, and Hautmann observed that the species with the more easily dissolved aragonite shells suffered a little bit more during the end-Triassic crisis. But, as noted in chapter 3, the rate of dissolution of both shell types is unlikely to be a serious issue in the lifetime of the bivalves.

Acidification and anoxia cannot be the whole story of the end-Triassic crisis. The hat-shaped radiolarians are unlikely to have been affected by either of these environmental factors—their silica skeletons would not been damaged by increased acidity, and the ocean's surface waters where they lived are unlikely to have suffered oxygen deprivation. Also, for life on land, something else must have done the damage. The nature of the abrupt extinction, with a short-lived fern spike, suggests a short-term crisis. Intense warming caused by peak volcanism would fit the bill, as would ozone depletion. As with the Siberian Traps, CAMP volcanism at its peak may have been substantial enough to release devastating concentrations of ozone-destroying halogen gases. Large-scale volcanism is also capable of generating very short cooling episodes caused by sulfate aerosol formation. Perhaps a brief but severe pulse of cooling, lasting only a decade or so (the sort of duration of volcanic cooling episodes), contributed to the extinctions? There is, however, no geological evidence to support this idea; all the data suggest rapid warming during the extinction. For now though, we cannot say for certain which of the volcanic gases—halogens or sulfur dioxide—were the deadliest for terrestrial life, only that they acted quickly.

Before leaving the end-Triassic crisis, it is worth mentioning the possible role of meteorite impact because there is some intriguing evidence. It may be recalled that the Rhaetian sediments of Britain have a seismite bed found immediately above the extinction horizon (see fig. 5.2). This bed can be traced across hundreds of kilometers, implying a huge earthquake that may have been, just possibly, caused by an impact. As it happens, a crater of approximately the right age has now been found near the town of Rochechouart in the Limousin region of western France. The crater has been much changed by subsequent tectonic movements, but its original diameter is estimated at around 40 kilometers. Sadly for proponents who like the death-by-meteorite-impact scenario, this crater size is not especially big. In fact, quite a lot of craters of this size are known. The Rochechouart impact is therefore unlikely to have caused global environmental catastrophe, as shown by comparison with other impact craters. For comparison, the Chicxulub crater associated with the dinosaurs' extinction is nearly 200 kilometers in diameter, indicating the size of impact required to be implicated in a mass extinction. Aside from Chicxulub, the largest meteorite impact crater is to be found in the Manicouagan region, in Quebec, and measures nearly 100 kilometers in diameter. At one time tentatively linked with the end-Triassic mass extinction, the Manicouagan crater is now known to have formed around 15 million years before the end-Triassic extinction, within the Norian Stage, a time noted for its lack of extinctions. The message here is that the threshold for lethal crater size needs to be in excess of 100 kilometers.

The Rochechouart impact is an unlikely cause of mass extinction, and it may not even have generated the British seismites. The onset of CAMP volcanism could have generated numerous violent earthquakes and shaken the seafloor sediments of the Rhaetian Sea. It thus seems more plausible that the seismic events record CAMP volcanism—the true killer of the Late Triassic. The consequence of these eruptions, like the Emeishan, Siberian, and Wrangelian eruptions before them, was a disaster for life and the environment. Recovery began immediately, as the dinosaurs and ammonites dusted themselves off and got on with business of taking over the world. But the bad times were not over yet. There was still another flood basalt province and another extinction to come.

CHAPTER 6

PANGEA'S FINAL BLOW

JURASSIC GOLGOTHA

The recovery from the end-Triassic catastrophe was in many ways a much swifter and more successful affair than that which followed the end-Permian debacle. A million years after the start of the Jurassic, seafloor life was both abundant and diverse. In contrast, a million years after the start of the Triassic, life was still in the doldrums—only a handful of small gastropod and bivalve species eked a living in the hot, oxygen-poor seas. However, Jurassic recovery was not all it seemed; many of the survivors were never to regain their former glory. Taking the ichthyosaurs as an example, many types went extinct and the few survivors never reevolved the same range of body types—only dolphin-like fish lizards were to grace the Jurassic seas. It would take a range of other newly evolved marine reptiles, such as the plesiosaurs, pliosaurs, and marine crocodiles, to add to the diversity of the ocean's top predators. The ammonoids reveal a very different response, although they too were reduced to a handful of species during the extinction. These gave rise to the ammonites, one of the most abundant and successful (and beautiful) of Jurassic fossils, but initially they were restricted to simple shells lacking ribs or fancy ornament—such embellishments would take a few million years to evolve.

Reefs are in many ways the most delicate of marine communities: highly prone to being wiped out during most mass extinctions with prolonged recovery times required in their aftermath. This was the case after the end-Triassic mass extinction. However, within 20 million years, corals were once again constructing reefs of appreciable diversity, but all this evolutionary effort was about to be undone by impending catastrophe. Trouble was about to strike in a familiar place.

The Karoo Basin of South Africa is doubly famous among the mass extinction fraternity; first, because it provides the best fossil record of the fate of terrestrial animals during the Permo-Triassic mass extinction as discussed in chapter 3, and second, because it was the site of vast flood basalt eruptions in the Early Jurassic that coincide with the final crisis in Pangea's history. The Karoo eruptions marked the onset of the continental rifting that was to ultimately break apart southern Pangea and create the continents that now occupy the southern hemisphere. As a result, as with the CAMP volcanics, formerly united lava piles are now separated between three continents (in this case Africa, Antarctica, and Australia), with the greater part being in South Africa. The Antarctic equivalents occur in Victoria Land, where the lavas are named after the Ferrar Glacier, which was in its turn named after the geologist Hartley Ferrar. The whole LIP is named the Karoo-Ferrar Province, and its original volume is estimated to be more than 3 million cubic kilometers of lava, making it a large example of its kind.

Like many LIPs, it was originally thought that Karoo-Ferrar eruptions spanned tens of millions of years, but dramatic improvements in dating precision in the 1990s and since have shown this to be a considerable overestimate. Recent work has shown that the lavas were erupted in a mere million years or so approximately 183 million years ago. This point in time is in the Toarcian, the last stage of the Early Jurassic, a period of major oceanic and climatic changes and significant extinctions.

The best evidence for the nature of the Early Jurassic crisis comes from a series of lovely coastal outcrops in North Yorkshire, northern England. These are centered on the fishing/tourist town of Whitby, which is famous for a Benedictine abbey that sits on the hill overlooking the harbor, and for its fossils, especially ammonites. The abbey and the ammonites are linked because St. Hilda, the seventh-century founder of the abbey, is said to have turned the local snakes into the coiled stone effigies (i.e., ammonites) that are found in the local cliffs and on the beaches. These are the targets of the numerous fossil hunters who visit the area, including me. I have many happy childhood memories of holidays on the Yorkshire coast, armed with a hammer and chisel and my mother's instructions to keep an eye on the tide. The latter is important because the tidal range of the North Sea is considerable, and it is easy to get cut off if you are not wary. However, this is to the benefit of geologists, paleontologists, and small children who like to play in rock pools because at times of low tide there are vast stretches of wave-cut platforms. During high tide one has to retreat to the amusement arcades and cafes in Whitby.

The Lower Jurassic cliffs around Whitby accumulated in a shallow sea that lay to the north of the Tethyan Ocean. In addition to ammonites, bivalves are very common, and many belong to long-lived species that first appeared at the start of the Jurassic and survived up to the start of the Toarcian Stage—a period of 20 million years. Deeper-water, dark-gray marine shales succeeded the shallow marine sediments at this time, and most of the long-lasting bivalves coincidentally disappeared—their time had finally come.

This extinction takes place in the first zone of the Toarcian, which is named after the ammonite *Dactylioceras tenuicostatum*. Tony Hallam was the first to draw attention to this increase of water depth coincident with extinction among bivalves, and he showed that the precise level of the losses coincides with a tough bed of black shale called the Jet Rock.[17] This bed marks the onset of a phase of anoxic deposition in the lower water column that was to persist for much of the Toarcian Stage. The reason for the bivalve extinctions is therefore immediately clear: in the absence of oxygen, the seabed became uninhabitable. Oxygen remained plentiful in the upper water column, though, as shown by the abundance of nektonic (free-swimming) fossils that lived there, including ammonites, fishes, marine reptiles, and a newly evolved group called the belemnites. These squid-like animals had an internal skeleton that included toward their back end a solid calcite structure shaped like a bullet and called a guard. Belemnite guards (usually just called belemnites) are prolifically common in the Toarcian shales, although the rest of the animal is rarely preserved, and as we shall see, the guards have provided a bonanza for geochemical study.

After Tony's pioneering work, one of my colleagues at Leeds, Crispin Little, looked in further detail at the timing of the Early Jurassic extinction losses, and he showed, along with his PhD supervisor, Mike Benton, that the crisis was not a single event but rather was spread over perhaps several million years. Further studies have resolved this further and shown that most extinction losses can be grouped into two distinct phases, especially among the ammonites. The main ammonite extinction occurs at the boundary between the Toarcian Stage and the earlier Pliensbachian Stage and thus was a little bit earlier than the bivalve losses, which were mainly late in the *D. tenuicostatum* Zone.

Around Whitby the change in ammonite faunas is very obvious to fossil collectors because thick-ribbed ammonites belonging to the amaltheid family are found in Pliensbachian strata and they abruptly give way, at the start of the Toarcian, to fine-ribbed dactylioceratids. These were immigrants. They had migrated from the warmer waters of the Tethyan Ocean to the south, and their appearance at Whitby can be read both as a refilling of empty habitats following the loss of the incumbents and as a response to increased warmth driving tropical faunas northward. The ammonite transition occurs precisely at the Pliensbachian-Toarcian boundary, which is marked by a thin bed of black shale called the Sulphur Band. The bed gets its name from its weathering appearance. It has a high content of pyrite, and this iron sulfide mineral decomposes readily to give a veneer of red iron oxides and the yellow of sulfur. The abundance of pyrite indicates that anoxia occurred during deposition of the Sulphur Band, and like the Jet Rock, this black shale is almost entirely devoid of fossils. Therefore both the slightly younger bivalve extinction and the main ammonite extinction coincide with anoxic conditions, although the Sulphur Band anoxia was of much briefer duration than that recorded by the Jet Rock.

Thanks to several detailed studies on this interval, we can date the sequence of Pliensbachian-Toarcian events with considerable precision. Thus the boundary between these two stages occurred a little before 183.5 million years ago, and the deposition of the Sulphur Band that followed, along with the coincidental first extinction pulse, probably lasted no more than a few thousand years. There was, then, a 300,000-year improvement in oxygenation before the next extinction and the onset of Jet Rock deposition. The second phase of intense oxygen starvation was of much longer duration; black shale deposition was to last 2 million years. As best we can tell, the onset of Karoo-Ferrar volcanism occurred early on in this prolonged anoxic episode, around 200,000 years after the mass extinction, a mismatch in timing that we will return to.

The losses recorded in the cliffs around Whitby were significant; many long-lived species disappeared, but it is difficult to claim that they were truly catastrophic in the same way as the Permo-Triassic and end-Triassic mass extinctions. The Toarcian extinctions did not cause wholesale changes in the composition of marine communities. Bivalves dominated on the seafloor, and ammonites and belemnites filled the water column both before and after this crisis. It is a similar story for other marine animals such as gastropods, forams, and crinoids—all suffered significant but not severe extinctions. The exceptions to this terrible-but-not-quite-a-catastrophe pattern are seen among the brachiopods and ostracods, two groups that faired very badly at this time.

The heyday of the brachiopods was before the Permo-Triassic mass extinction, and they never recovered their previous dominance following the disastrous losses incurred during this crisis. However, by the Early Jurassic they had attained a respectable diversity and were common in the warm waters of the Tethyan Ocean but less abundant in the shallow seas that covered England. Brachiopods show distinct changes in the early Toarcian: like the ammonites, many species migrated northward from Tethys into the seas covering southern England, suggesting a warming trend (discussed further, below), but their main story was one of extinction. Two entire orders disappeared at the end of the Pliensbachian—the Spiriferinida and Athyridida—bringing to an end two dynasties that had lasted for more than 250 million years. It is fair to say that the brachiopods were truly knocked for six[18] by this Jurassic devastation: more than three-quarters of their genera and perhaps 90% of their species failed to survive. No new orders of brachiopods arose to replace the losses, and only two orders are common today—a minor memory of their Paleozoic glory days.

For the ostracods (tiny crustaceans that live encased inside two bean-shaped shells), the Toarcian crisis was even worse; an entire suborder, the Metacopina, was lost—the highest-ranking taxonomic extinction of the entire crisis—but their recovery was much more successful. Today ostracods are a common and diverse component of aquatic communities.

Beyond the shallow shelf seas of northwestern Europe, the Toarcian crisis also wrought its changes. One of the strangest groups to succumb was an unusual group of bivalves called the lithiotids. Throughout most of their history, bivalves evolved at a rather slow pace and generally restricted themselves to a limited range of forms familiar to anyone who has gone shell hunting on a beach. However, every so often large, thick-shelled forms with strange morphologies have appeared. We have already encountered one family, the Triassic megalodontids, which were victims of the end-Triassic mass extinction, and in the Early Jurassic the lithiotids were another evanescent group to appear in the shallow seas fringing Tethys. They were denied a longer history by the Toarcian crisis. I have studied their extinction in limestones in southern Tibet, where they disappear from a series of beds that record deepening. In this regard, their extinction is comparable with that in northern Europe, but the Tibetan extinction/deepening is not associated with the onset of anoxic, black shale deposition. Indeed, marine anoxia seems to have had a very patchy development in the seas of the southern Tethyan shelf.

Turning our attention to the Panthalassa Ocean, we can examine the fate of the planktonic radiolarians. The best evidence, like that for all previous Pangean crises, comes from chert beds that formed on the Panthalassa Ocean floor but are now found in the mountains of Honshu, central Japan. The cherts reveal a sharp decline in the variety of radiolarians during the Pliensbachian-Toarcian and an early Toarcian population that was characterized by relatively few species of spumellarians—one of the morphologically simple group of radiolarians. So, clearly something was happening to plankton populations in the oceans, but the changes proved to be short lived. There were no catastrophic losses for radiolarians on the scale of those in the Permo-Triassic wipeout or even the end-Triassic extinction. Instead, immediately after the spumellarian bloom, most of the radiolarian groups that had gone reappeared once more.

The clearest evidence of a Toarcian oceanic crisis comes not from the radiolarians but rather from the sediments in which they occur because a thin horizon of organic-rich shale is interbedded among the Japanese cherts. Very similar to (and contemporaneous with) the Jet Rock in Whitby, this bed also records anoxic seafloor conditions, although in this case, the oxygen starvation was developed far out in the open ocean. The Japanese black shales contain tiny pyrite framboids with the same size ranges of those that formed in the Early Triassic (see p. 57), and so like the earlier times, conditions of intense anoxia with hydrogen sulfide present in the water column had returned to Panthalassa. All told, this Toarcian oceanic anoxic event, as it has been termed by geologists, has a somewhat hit-and-miss development: although clearly expressed in European shelf seas and the open ocean, it has a very muted development in the waters of Tethys.

The radiolarians are not the only plankton that left a good fossil record in the Early Jurassic. As noted in chapter 4, the coccolithophores are a group of plankton that secrete tiny calcitic plates, called coccoliths, that form deep-sea carbonate mud. They appeared in the Late Triassic, and by the end of this period they had diversified somewhat, only to loose significant numbers of species in the mass extinction. However, they bounced back successfully and began to proliferate, initially in the shallow seas fringing Tethys and later in the open ocean waters of Tethys. At the time of the early Toarcian crisis, they were still mostly restricted to shelf seas. How did they fair at this time? Surprisingly the answer is, really quite well. Compilations of diversity show no hiccup; instead, the coccoliths' increasing diversity continued unchecked until the end of the Cretaceous, nearly 120 million years later. The Toarcian crisis clearly did not have a long-lasting impact on the world's plankton.

Having surveyed the changes in the Jurassic oceans, let us now turn our attention to the land. By the Early Jurassic, the dinosaurs had taken over and had diversified into forms that included large, armored herbivores; even larger, long-necked sauropods; and predatory theropods, such as *Dilophosaurus*. The last makes an appearance in the first *Jurassic Park* movie, where it has a nasty spitting habit. Unfortunately though, our knowledge of dinosaur fortunes around the time of the Toarcian crisis is not very good, which may have to do with the very high sea level at this time, making terrestrial dinosaur-bearing sediments rather rare. This lack makes it difficult to tell if there were any significant losses among the dinosaurs. There is a little bit of indirect evidence that suggests something may have happened. Matthew Carrano of the Smithsonian Institution undertook a huge, exhaustive study of the evolutionary tree of the megalosaurids—giant predatory dinosaurs—which showed that they apparently underwent a major diversification in the later Toarcian. Such radiations often follow on from mass extinctions because they remove incumbents and thus provide an impetus for evolution to re-fill vacated niches, so perhaps a dinosaur extinction event in the early Toarcian removed the existing predators like *Dilophosaurus* and allowed the megalosaurids to step forward and take their place. Unfortunately, evidence like this is no more than circumstantial. Not every evolutionary success story has to be preceded by extinction. It would be fair to say that had there been a truly major terrestrial extinction in the Toarcian, then paleontologists would probably have found the evidence by now.

Despite the patchy effects of the early Toarcian crisis on the world's ecosystems, it is clear that the oceanographic changes were spectacular, especially the development of a phase of oceanic anoxia comparable in intensity with that of the Early Triassic oceans. There is also plenty of geochemical evidence for major changes in the atmosphere. Much of this support comes from belemnites. These are very solid, chunky bits of calcite that have altered little since they were originally grown, so the constituent carbon and oxygen atoms of the calcite retain the original isotopic ratios of the Jurassic seawater in which they formed. This is doubly fortunate because hitherto, in our study of the Pangean world, conodonts have been the fossils of choice when compiling isotopic trends. As we saw earlier, these animals disappeared forever during the end-Triassic mass extinction, to the great disappointment of geochemists (and no doubt the conodonts), but fortunately the belemnites stepped in to fill a vacancy in the category of incredibly useful fossils.

Carbon isotope measurements from belemnite calcite show a shift to heavier ratios (i.e., there is more of the carbon-13 atom) in the early Toarcian at the time of deposition of the Jet Rock and Panthalassa Ocean black shales. This change makes a great deal of sense because the abundant organic matter in these black shales is rich in carbon-12. We would expect the removal of a lot of light carbon from the oceans (in the form of the shales) to leave the remaining carbon enriched in heavy carbon, which is what the belemnite calcite carbon isotope ratios show. All well and good then, but we also have another way of measuring the isotopic record of carbon in the oceans and atmosphere, and that is by analyzing organic carbon. Remarkably, this carbon does not show the same trend as in the belemnites, even though it is coming from the same source—the atmosphere and oceans of the Early Jurassic world. The shift to the heavier isotope is still recorded in the organic carbon isotopes, but it is preceded by a major swing to values enriched in carbon-12 around the time of the Toarcian extinction. This earlier negative shift is very similar to that seen around the onset of all the other mass extinctions discussed in this book, and so we should perhaps not be surprised to find it happening again during the Toarcian crisis. Needless to say, the discrepancy between the belemnite and organic carbon isotopic values has caused a huge amount of (sometimes acrimonious) discussion among geologists— which need not concern us here—but finding the reason behind it has so far proved elusive.

If the negative shift in organic carbon isotope values is genuine, then it may record a major influx of greenhouse gases into the atmosphere as seen during other extinction events (see p. 62, for example), in which case we should find evidence for global warming. This prediction is supported by such fossil distributions as the northward migration of warm water brachiopods and ammonites from the Tethyan Ocean. It is also supported by belemnite analyses (again!), but this time of their oxygen isotope ratios. A little over ten years ago, John McArthur (University College London) and his colleagues were the first to produce a temperature curve using the oxygen isotope values from Whitby belemnites. These ratios showed that the onset of anoxic conditions, seen in the Jet Rock and other black shales, coincided with a rapid rise in seawater temperatures of up to 7°C. John speculated that the two phenomena were related as follows. Rising temperatures typically go hand in hand with increased humidity, in turn causing increased runoff from the continents (because of the extra rainfall) and thus an extra supply of terrestrial nutrients into the sea. These nutrients are food for plankton, which therefore increase in abundance. The decay of all this extra planktonic organic matter in the oceans would raise the rate at which oxygen is depleted, and voilà, you have a link between warming and anoxia. This link between nutrient supply to the oceans and anoxia has also been proposed as one of the causes of Permo-Triassic ocean anoxia, in which we saw that the links between warming and anoxia are manifold.

Seawater temperatures derived from oxygen isotopes can also be compared with evidence for warming derived from analysis of leaf stomata density. Once again Jenny McElwain has done the hard work, and her results are rather intriguing. Samples of fossil leaves from the Danish island of Bornholm revealed a rapid rise in atmospheric CO_2 from 600 ppm in the middle of the *D. tenuicostatum* Zone to around 1000 ppm at the end of this interval—a perfect fit with the oxygen isotope trend. However, after a gap in the fossil record, the next set of leaves shows that atmospheric CO_2 values plunged to 350 ppm, implying a sharp cold snap. This result does not fit with the belemnite oxygen isotope trend at all because this suggests the acme of warming occurred at the same time as the leaves record cooling. I suspect the stomatal data is at fault in this case and has been overinterpreted, because the published data is "noisy" during the so-called cool phase.

Further refinements to the Jurassic seawater temperature curve have been made using samples taken from Spanish locations. These have reproduced the original temperature curve but also include an additional, brief warm "blip" at the start of the *D. tenuicostatum* Zone. This neatly coincides with the first pulse of extinction (particularly of ammonites) and also the black shale that developed at this time. It seems as if the beginning of the Toarcian witnessed the first attempt to move to a world with warm climates and anoxic seas, but feedback mechanisms swung conditions back to cooler, better-ventilated seas. Nonetheless, the Pliensbachian-Toarcian boundary changes were sufficiently bad to cause problems for ammonites. Only later in the *D. tenuicostatum* Zone was there a more permanent shift to warmth and anoxia, and this time the extinctions were concentrated among seafloor creatures.

With these temperatures trends established, it becomes clear just how remarkably similar the Toarcian crisis was to the Permo-Triassic mass extinction. They share the following attributes:

1. Two main phases of marine extinctions separated by an interlude lasting 200,000 to 300,000 years that allowed a temporary respite and recovery
2. Coincidence between rapid warming of oceans and atmosphere and sea-level rise
3. A first extinction phase that coincides with a short anoxic interval and a second extinction phase that occurs at the start of a much longer phase of ocean anoxia lasting a few million years
4. Carbon isotope trends that become lighter (more carbon-12 relative to carbon-13), suggesting a lot of light carbon is being added to the atmosphere

And of course there is also the correlation with the eruption of a LIP, although in the case of the Karoo-Ferrar eruptions these seem to have begun 200,000 years after the second extinction—they were a little bit late. In fact, this slight mismatch in timing is also seen for the Emeishan Traps, Siberian Traps, and CAMP eruptions. The cause of the discrepancy probably lies in way the LIPs are being dated. Working out the ages of volcanic rocks involves extracting crystals from lavas and finding ones that are rich in radioactive elements. Analysis of these allows an age to be determined. There is nothing wrong with this approach—it gives the age of the main interval of eruptions—but it is likely that all the environmental damage was done by gigantic gas eruptions right at the onset of the eruption history before the main lava flows appeared. The reasons for this lie in the nature of the mantle plumes that are the source of LIP magmas: it is probable that a lot of volatile, gas-rich eruptions preceded the main lava flows.

Direct evidence in support of the giant gas eruptions can be found in the Karoo landscape of South Africa. The region is peppered with thousands of small hills called volcanic breccia, cones that measure around 100 meters in diameter and 10 to 20 meters high. There are no less than 430 of them around the small town of Loeriesfontein alone. Below the surface, the cones are connected to narrow shafts of broken-up rock that formed by the explosive release of gas; these are called breccia pipes. The magma that fed the lavas in this region ascended through continental crust that has thick layers of Permian black shales. As this hot material came into contact with the shales, it probably would have baked them and released huge volumes of CO_2—suggested to be as much as 2000 gigatonnes of CO_2. Because the carbon dioxide was derived from isotopically light organic matter in the shales, then it too would have had light isotopic values. It thus likely that the negative carbon isotope excursions seen in contemporary organic carbon is recording this major episode of carbon dioxide release. This is, of course, Svensen's hypothesis all over again. It has been a regular feature in all of the proposed extinction scenarios of Pangea.

Putting all these observations and ideas together it is possible to come up with an extinction scenario that is essentially the same as that proposed for all the other Pangean extinctions, especially the Permo-Triassic crisis (fig. 6.1). A sudden eruption of flood basalts in southern Pangea released great volumes of CO_2 and thereby triggered a warming trend. Additional CO_2 from crustal baking (Svensen's hypothesis) may have further contributed to this heating. Once established, the warming trend may also have initiated the release of methane from gas hydrates, although this aspect is not a compulsory feature of the kill model; the gases linked with volcanism may have been sufficient.

A point was rapidly reached at the start of the eruptions where the warming trend caused marine anoxia to become

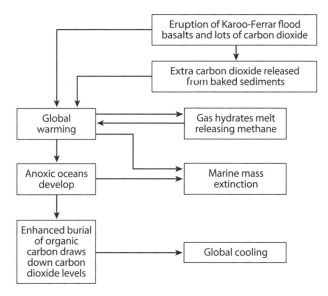

Figure 6.1. Chain of cause and effect for Toarcian Age (Early Jurassic Period) environmental change and extinction crisis.

widespread (for the link between warming and anoxia, see the discussion in chapter 3, pp. 76–79). This set up negative feedback that gradually returned the world to less harsh conditions. The mechanism works as follows: anoxic conditions favor the preservation of organic matter, which causes black, organic-rich shales to accumulate. The carbon in black shales is derived from phytoplankton, which have removed carbon dioxide as they photosynthesize. Thus, organic burial is a highly effective way of removing atmospheric CO_2 and reversing greenhouse trends. The shift to heavier carbon isotopes in the seawater at the time of black shale formation is evidence of this process occurring because isotopically lighter carbon is being buried in the black shales.

Intriguingly, as with the Permo-Triassic crisis, it would appear that this cycle of warming anoxia and cooling happened at least twice: once (briefly) at the Pliensbachian-Toarcian boundary, and across a longer period beginning in the late *D. tenuicostatum* Zone. The implication is that there were two blasts of gas from the Karoo-Ferrar LIP.

The direct cause of extinction in this Toarcian scenario is generally thought to be the spread of anoxic conditions and the consequent loss of habitable seabed environments and water columns for marine life. The link is especially clear in northern Europe, where black shales are well developed, but less so in southern Tethys, where the oxygen deficiency was less intense but perhaps still sufficiently damaging. Were high temperatures also involved in the extinctions? Several geologists are certainly keen on this idea as a factor in the brachiopod extinction, although the temperature rise does not seem to be that extreme. Temperatures, calculated from belemnite calcite data, suggest a rise in Spanish seas from 13°C to 21°C during the extinction. If these values are reliably accurate, they are not likely to have been detrimental to marine life; indeed, they are comparable to those around the Mediterranean shores of Spain today. It is difficult to imagine such temperatures playing any role in the Toarcian crises.

As with the earlier mass extinctions of Pangea, the rapid release of carbon dioxide to the atmosphere can lead to ocean acidification and thereby inflict a further potential stress on marine life, especially among shelled invertebrates. Some evidence for this comes from the study of coccoliths in western Tethys. Emanuela Mattioli of the University of Lyon has spent a career of examining coccolith fossils—something that requires a great deal of patience, because they are very small (rarely more than 10 microns in size), and a keen eye because, to nonspecialists at least, they all look rather similar. By counting their abundance in sediments from Tethyan shelf sediments, Emanuela has shown that the coccolith flux rate (how fast they were raining down to the seabed) declined in the early Toarcian as the coccoliths become smaller and lighter. It is tempting to ascribe these changes to the increased difficulty of secreting calcite in more acidic surface waters; however, other environmental factors could be at play. Coccolithophores do best in regions of low nutrient supply because they get outcompeted by other types of plankton when nutrients are abundant. The poor showing of the coccoliths during the Toarcian may therefore be more a signal of increased nutrient runoff in a warmer world rather than of a lower pH. Wisely, Emanuela has been careful not to make too emphatic a link between her observations and acidification.

Whatever was happening with the plankton in the Toarcian oceans, it was not a time of devastating crisis. Neither do the terrestrial animals appear to have suffered badly at this time. One can speculate on the reasons behind this restrained response to the Karoo-Ferrar eruptions. Marine anoxia was widespread but not anywhere near as extensive as that seen during the Permo-Triassic event, so there was plenty of seafloor not affected by these hostile conditions. Temperatures rose and no doubt were a key factor behind the anoxia, but values do not appear to have reached the same stressful levels seen during earlier crises.

All told, the final crisis of Pangea was a subdued, muted version of those that had gone before. It is almost as if life was finally getting used to the effects of LIP volcanism. Whether true or not, there is no chance to test this notion because for the remainder of the Jurassic, no more LIPs formed. Instead Pangea began to slowly drift apart as the Central Atlantic opened, and its southern arm, Gondwana, began to split up. Not until 40 million years after the Karoo-Ferrar eruptions did LIP volcanism once again return, and by this time Pangea had separated. This was the first of many giant flood basalt outpourings to form in the Cretaceous Period. And what happened to life at this time? This is a topic for the final chapter.

CHAPTER 7

PANGEA'S DEATH AND THE RISE OF RESILIENCE

Around 135 million years ago, giant cracks started to open up among the sand dunes found in a vast desert that lay at the heart of Gondwana. Soon magma welled up through these fissures, and the dunes were drowned in flows of basalt lava hundreds of meters thick. The lavas continued to pour out for hundreds of thousand years, ultimately to form one of the largest of all LIPs—the Paraná-Etendeka Province. Today these lavas are separated by the South Atlantic, with the larger part occurring in South America (Brazil and Argentina), where the Paraná River cuts through them, while a somewhat smaller area is found in the Etendeka Desert of Namibia. Like the earlier CAMP and Karoo-Ferrar provinces, this latest LIP marked the start of continental breakup that was, in this case, to lead to the separation of the South American and African continents. But in the context of this book, the key question is, did the eruptions lead to mass extinction and catastrophic environmental change like the preceding five LIPs of Pangea?

The answer is an emphatic and simple no. The Paraná-Etendeka eruptions occurred during the middle of the Hauterivian, the third stage of the Early Cretaceous, when global extinction rates were extremely low. For lovers of catastrophes, the interval is a distinctly dull one. Perhaps the only interesting event was the temporary loss of shallow-water limestones in the western Tethys. They were replaced with shales as the seas deepened. Despite this regional pause in carbonate formation, overall it was a good time for limestone because planktonic forams began to radiate and their calcite shells had begun to rain down on the seabed in huge numbers. Hitherto, forams had only lived on the seafloor, where they were (and still are) very successful despite the various extinction crises they suffered during Pangean time. Around the middle of the Jurassic, some of them began to spend their lives drifting in the surface waters of the oceans, but it was not until the Early Cretaceous that this mode of life became common.

Planktonic forams were not the only calcareous plankton in the Cretaceous oceans. The coccoliths, which we have already met, made their appearance in the Triassic. By the Cretaceous their range had expanded out from Tethyan shelf seas into the open ocean. By the Hauterivian the combined rain of coccolith and foram skeletons to the seabed was forming a new type of sediment—deep-water pelagic carbonate, which was composed entirely of these tiny microfossils. We will return to the significance of this lime blanket later.

By the Cretaceous it seems the LIPs had lost their teeth. Pangea had seen 80 million years of LIP-driven disasters, but the next 80 million years saw amazing, uninterrupted evolutionary success stories. Feathered dinosaurs began to fly, and recent spectacular finds from China have shown that by the Early Cretaceous this group, the birds, had already attained an impressive diversity of forms. Plants were even more productive during this interval. The Early Cretaceous saw the rapid rise of the flowering plants (angiosperms), which finally brought some new colors to the green terrestrial landscapes. The rise of the social insects, such as termites, ants, and wasps, at the same time may owe something to the ascent of the angiosperms. Meantime, in the oceans, the burgeoning of life was similarly unimpeded. Especially successful were the decapod crustaceans, which produced two new groups—the crabs and lobsters. Sea urchins also underwent some remarkable changes. Hitherto they had lived on the seafloor encased in an armored sphere with defensive spines. This type still thrives today, but in the later Jurassic some started to burrow deep into the sediment, producing highly modified forms known as heart urchins.

All this innovation and diversification could have been brought to an abrupt halt by the Paraná-Etendeka eruptions. Groups as diverse as the birds and heart urchins could have had a brief flowering and so be as obscure today as earlier mass extinction victims such as the megalodontids or crurotarsans of the Triassic. But this did not happen and there was no crisis.

The benign nature of the Early Cretaceous volcanism is starting to attract attention from geologists and may have something to do with where it occurred. Earlier Pangean LIPs are thought to have intruded crust full of "juicy" rocks, such as coals and black shales, that when baked by magma generated huge volumes of greenhouse gases. This is the Svensen Hypothesis, and as we have seen, it has been deployed as a partial explanation for all the crises of Pangea. The crust beneath the Paraná lavas, on the other hand, is old and dry; it consists of already metamorphosed rocks that will not release many gases if baked by magma.

Direct evidence for the gas-poor content of the crust beneath the Paraná-Etendeka lavas has recently come from studying the sulfur content of pyroxene crystals in basalt. Pyroxene is one of the most abundant crystals found in LIP igneous rocks, and although no sulfur is found within its crystal lattice, it is possible for tiny inclusions or impurities to be trapped within the crystal. Analyzing them provides the chemical composition of the magma from which the pyroxenes crystallized. Thus, the Paraná-Etendeka magmas are thought to have contained around 800 ppm sulfur, whereas the CAMP magmas had over three times as much—1900 ppm. Perhaps this was the reason for the failed environmental response? There was simply not enough gas released to do any harm. This interesting notion is testable (and can be shown to be wrong) because the Paraná-Etendeka eruptions were just the first of a whole series of LIPs that formed since the start of the Cretaceous. Some of them passed through some very gas-rich crust indeed, especially the North Atlantic Igneous Province, which erupted 75 million years later. The effects of these younger eruptions are described further below, but they are not linked with a mass extinction.

LIPS AND LIFE

After the Jurassic Period, it becomes a lot easier to study the past environmental history of Earth because a lot of ocean crust is preserved from this time onward, whereas all the older crust has been lost through subduction back into the mantle. Remember that in previous chapters we had to look at much-deformed ocean-floor sediment in places like Japan to piece together what happened in the Panthalassa Ocean—a very imperfect and fragmentary record analogous to a jigsaw puzzle with most of the pieces missing. In contrast, the Cretaceous ocean floor puzzle is mostly complete because a lot of it is still at the bottom of the ocean, and numerous international ocean drilling programs have provided many cores through it. One noteworthy feature of Cretaceous oceans is that there are several giant plateaus. They are composed of sheets of flood basalt that formed from plume eruptions, just like the continental flood basalt provinces, but this time under the ocean. We have already met one example, the Wrangelia Province of western Canada, which formed within the Panathalassa Ocean at the time of the Carnian crisis, although it is now part of the Canadian Rockies (see chapter 4). Several of the Cretaceous ocean plateau eruptions coincide with interesting environmental changes that typically include such familiar themes as global warming and ocean anoxia.

The Ontong Java Plateau, which is found in the southwestern Pacific, is the largest of all the oceanic flood basalt regions. It is not especially well dated—because it is largely inaccessible beneath the waters of the Pacific—but it probably formed some time between 125 and 118 million years ago. Within this interval there is an interesting, short-lived oceanic phenomenon called the Selli Event, which happened 123 million years ago. It is marked by the spread of black shales in the deeper parts of the Pacific and Atlantic Oceans, a big change in carbon isotope ratios, and a warmer climate. All of this sounds very familiar. The changes bear close comparison with some of those seen during the Pangean crises, especially the Toarcian extinction. However, the Selli Event did not coincide with increased extinction rates, although a few groups show some interesting changes. One such is the rudists, a family of large thick-shelled bivalves similar to the megalodontids of the Triassic and the lithiotids of the Jurassic, both of which succumbed during Pangean extinctions. Many rudists disappeared from the fossil record during the Selli Event, but only temporarily, because they reappear after the event. The Selli Event was thus an interval of interesting environmental changes that never became a crisis for life.

The next great oceanic anoxic event, 29 million years later, was a similarly fascinating but relatively harmless episode. Called the Bonarelli Event, it too coincided with a hot climate, carbon isotope changes, and the eruption of oceanic flood basalts, this time in the Caribbean and Columbian regions. There were a few extinctions, including some planktonic foram species, with the result that a mass extinction epithet is sometimes used for this interval,[19] but by the standards of Pangean disasters, it was a very minor crisis.

The possible standout exception to the harmless LIPs of the past 135 million years is the Deccan Traps—the product of a vast outpouring of flood basalts in western India 66 million years ago. They famously erupted at the moment that dinosaurs and numerous other groups, including the ammonites, marine reptiles, and nearly all species of calcareous plankton, went extinct. This is known as the Cretaceous-Tertiary mass extinction, the subject of some of the most vociferous scientific arguments of the past forty years because, as surely everyone knows, the extinction also coincided with a giant meteorite impact at Chicxulub. So we have a choice: either the meteorite did it or the lava did it or it was some combination of these two. Many books have been written about the Cretaceous-Tertiary extinction and what caused it, but this is not the place for an exhaustive analysis. Fortunately though, the huge efforts that have gone into the study of the disaster provide a fantastic temporal resolution, especially the climate changes leading up to the mass extinction, and they repay some consideration.

In the million or so years prior to the Cretaceous-Tertiary mass extinction, there was a long-term cooling trend that sharply reversed to warmer conditions around half a million years before the end of the Cretaceous. The warm interval was then halted by a sharp cooling about 50,000 years before the end of the Cretaceous. Analysis of fossil plants in North America suggests that the cooling was substantial, maybe as much as 8°C, and was followed by stable temperatures across the boundary itself. So how does this compare with the eruption history of the Deccan Traps? The eruptions began about half a million years before the Cretaceous-Tertiary boundary, thus suggesting that the LIP may have initiated a warming climate—a link also seen during the Pangean LIP eruptions. But what about the subsequent cooling trend? Was it triggered by an especially explosive phase of volcanism injecting volcanic gases high into the stratosphere? Alternatively, did all the freshly erupted Deccan lavas start weathering intensely, thereby removing carbon dioxide from the atmosphere and cooling the planet? It could be either of these alternatives, and both have been suggested as explanations, but what is clear is that these potentially Deccan-driven climate changes predate the mass extinction at the boundary. The precise temporal link between meteorite impact and extinctions has always been the strongest argument in the debate on the cause of the mass extinction. Nonetheless, the debate rumbles on and there are those who claim the impact at Chicxulub was thousands of years before the mass extinction.

Latest Cretaceous extinctions are known, especially among North American dinosaurs, and they are often used as evidence that the world was in (volcanic) trouble even before the meteorite struck. However, it is also possible that the dinosaur losses are a regional climate signal. Remember, there was a transition from warm to cool conditions 50,000 years before the end of the Cretaceous. Diversity declines with temperature, as is immediately apparent if you consider the diversity of life in the Amazon compared to the Arctic tundra. Thus a cooling trend will see the loss of diverse, warmth-loving communities and replacement with less diverse, cold-loving communities. This local decline in diversity does not have to mean that there was a change of global diversity. If cooling was the cause for North America diversity decline, then it becomes hard to argue that the onset of the Deccan eruptions caused any significant extinction.

Disentangling the role of the Deccan eruptions and the meteorite impact in the Cretaceous-Tertiary mass extinction is clearly tricky, but fortunately the next LIP eruption presents a much "cleaner" story uncomplicated by a coincidental meteorite impact. This eruption occurred only 6 million years after the dinosaurs' demise, and it triggered the onset of rifting between the North America and European-Asian continents. It was, in a sense, the final fragmentation of Pangea. Today the lavas are split between Greenland and the British Isles, with Iceland recording the remnant volcanism of the plume that fed the lavas. The province is termed the North Atlantic Igneous Province (NAIP), and the volume of lava extruded was truly huge even by LIP standards, probably more than 6 million cubic kilometers. In addition, the magma passed through some very organic-rich sedimentary rocks, and it seems likely that vast volumes of gas were emitted from baked sediments together with the volcanic gases. Indeed, Henrik Svensen devised his original hypothesis—of enormous thermogenic gas release from LIPs—based on a study of NAIP lavas in the North Sea.

Giant amounts of volcanism and giant gas release—if anything could replicate a Pangean LIP catastrophe, then the NAIP is clearly the best candidate to do so. So did it? The answer depends on whom you read, but there were certainly some fascinating environmental changes around the time of eruption, most significantly 70,000 years of intense global warming. That period occurred at the Paleocene-Eocene boundary, which has led the climatic episode to be called the Paleocene-Eocene thermal maximum (PETM). It coincides with a loss of ventilation in the deep-ocean waters of the Atlantic, where oxygen-poor conditions developed, and a sharp shift in carbon isotope ratios that become 3 ‰ lighter. Once again we have the ingredients of a Pangean-style mass extinction, and so we might expect to encounter one at this time.

Many have indeed argued that this was the case; for example, one such doom-laden analysis in a popular science magazine proclaimed that during the PETM

> the earth warmed faster than at almost any other time in its history. The average temperature soared by 9 degrees Fahrenheit, entire ecosystems shifted to higher latitudes, and massive extinctions occurred on land and, most telling, at sea. James Zachos, a paleoceanographer from the University of California at Santa Cruz [says] bottom-dwelling creatures with shells disappeared from the fossil record . . . [and] it took 60,000 years before sediments again began to show a thick white streak of fossilized shells.[20]

This sounds truly terrible, almost as bad as the Permo-Triassic mass extinction; however, the assessment of the severity of the crisis is untrue. Extinction losses were very selective and very minor during the PETM. Some deep-sea forams and ostracods in the Atlantic Ocean went extinct, probably because bottom waters became oxygen poor, while a few larger forams in tropical shelf seas also disappeared, but on the whole, this was a time of evolutionary success for a broad range of animals. Mammals did especially well. Many modern families appeared, including our own group, the primates, together with the first members of the deer and horse families. Interestingly though, most mammals were exceptionally small during the PETM. I suspect this was Bergman's rule in action (see p. 98), according to which high temperatures favor small organisms.

Part of the problem for PETM scientists who claim that there was a crisis is that they are dealing with a spectacular climate event, in particular, a story of rapid warming that was probably driven by the release of greenhouse gases, and so an extinction is anticipated (just as a pending mass extinction is predicted for modern greenhouse warming). And yet, the PETM crisis was only the faintest echo of the Pangean extinctions. Claims to the contrary can be dismissed as mass-extinction envy on the part of those who study the PETM. What actually happened at this time is much more subtle and nuanced.

The best evidence for oceanic changes during the PETM comes from calcareous plankton. Samantha Gibbs (Southampton University) and Heather Stoll (Oviedo University) have pieced together the story and shown that the onset of the PETM is marked by the loss of pelagic limestones in some parts of the deep ocean. This loss does not reflect the extinction of the coccoliths and planktonic forams that form such sediments, because their diversity and abundance in surface waters remained unchanged, but rather the dissolution of the deep-sea carbonate sediment. The cause appears to have been extra organic matter raining down to the seabed, where it decomposed and acidified the sediment. The reason for the extra organics relates to the feedback mechanisms of the ocean-atmosphere system. Global warming increases rainfall, which flushes more nutrients into the ocean, thus stimulating more organic-matter production by plankton. This is the same old familiar process thought to have happened during the Toarcian and Permo-Triassic crises except that this time, during the PETM, there was no long-term effect.

Gibbs and Stoll also showed that the coccoliths responded in a perhaps surprising way during the PETM. Global warming driven by increased atmospheric CO_2 will also see a decrease in pH as more of this gas dissolves in the surface waters. This acidification should be to the detriment of coccolithophorids with their calcitic shell, but in fact they thrive better because, despite the subtle reduction in pH, the extra CO_2 stimulates more photosynthesis. This idea is supported by observations on modern coccolithophorids, which are found to grow thicker skeletons as carbon dioxide levels increase in sunlit waters. In essence, coccoliths provide a responsive feedback system for increased atmospheric CO_2 levels because they help mop up the extra carbon. Ultimately this ends up either as limestones on the seafloor or, in the deepest parts of the ocean where coccoliths dissolve, their carbon becomes part of the ocean's dissolved carbon content—another key process in lessening the effects of acidification.

So, is coccolith formation the key, new factor in helping moderate the effects of rapid carbon dioxide increases? Consider the following reaction for limestone formation:

$$Ca^{2+} + 2HCO_3^- = CaCO_3 + H_2O + CO_2$$

This shows that the formation of the calcium carbonate also produces two by-products—water and, problematically, carbon dioxide. Ostensibly this should exacerbate the amount of CO_2 in the ocean. However, two biocarbonate ions are used on the other side of the equation, and these are ultimately derived from rock weathering. Acid rain falling on the land surface attacks rocks and produces the bicarbonate ion, which gets flushed into the oceans. The ions increase ocean alkalinity and so help more carbonate to be precipitated. Overall, the two molecules of carbon in bicarbonate (which was originally found in two molecules of atmospheric carbon dioxide) are used up but only one is re-released as carbon dioxide.

The only problem is that the weathering process responds on a time scale of thousands of years, whereas a sudden increase of carbon dioxide emissions from, say, volcanism may occur in decades. So atmospheric carbon dioxide, along with surface water acidity, can increase quickly, whereas responses such as increased weathering and carbonate formation take thousands of years. The dissolution of deep-sea carbonate is another effective way of decreasing ocean acidity but, again, this is also a relatively slow process, because it takes hundreds to a few thousand years for ocean currents to mix the surface waters into the deeper parts of the ocean. These problems are of course at the nub of modern concerns of human-driven atmospheric CO_2 increases. Ultimately, much of the carbon dioxide we put into the atmosphere will end up in the ocean either as organic carbon, carbonate, or dissolved carbon compounds; it is just that effects like surface water acidification occur very quickly whereas the negative feedback mechanisms are a little bit slower. Fortunately, coccolithophorids help matters considerably because rather than release carbon dioxide during shell formation, it gets used during photosynthesis.

If coccoliths are the key factor in the improved response of earth's system to carbon dioxide increases, then the link between Pangea and LIP-extinctions may be entirely coincidental. The burgeoning of calcareous plankton and deep-sea limestone deposition may be the salient issue, not the absence of a supercontinent. This notion can be tested (sort of), by a quick survey of the extinction crises that occurred before Pangea assembled and before the rise of coccoliths. We need to look at the extinctions of the pre-Permian world.

A QUICK LOOK BACK

There was no shortage of extinction crises before Pangea although none were on the scale of the Permo-Triassic and end-Triassic cataclysms. Following the initial dramatic radiation of life, during the Cambrian explosion, the next 200 million years saw nearly a dozen severe crises, of which two merit the "mass extinction" epithet. Of these dozen, none are clearly linked with LIP eruptions although two may be. This weak link may reflect a problem with studying such deep time: these extinctions occurred a very long time ago, and many

LIPs from this interval have since been eroded. Often only the deep "plumbing system"—the giant dikes and sills that sourced the magma—remain, whereas the surface lavas have been stripped off. This makes it hard to work out the original size of the volcanic provinces, and their ages also become unclear because the igneous rocks of the plumbing system need not be exactly the same age as the erupted lavas. Herein lies another problem: dating igneous rocks gets less precise the further back in time you travel. Thus, eruption ages for lavas from more than 500 million years ago may come with errors of plus or minus 10 million years.

One of the strongest connections between pre-Pangean extinction and LIPs comes from the Cambrian Period. The Kalkarindji Province, a much eroded but probably originally, large flood basalt region in northern Australia, was erupted around 506 million years ago. This about the same age as the first mass extinction. The principal victims were the archaeocyathids, a group of heavily calcified sponges that constructed large, highly diverse reefs in the Early Cambrian. They disappeared abruptly just prior to the Middle Cambrian, and several geologists have blamed the Kalkarindji eruptions for their demise. Not long after this extinction, there was another one, and this time it eliminated the redlichiids. These were the first major family of trilobites to appear, and they were also the first to disappear, suffering a mass extinction at the boundary of the Early and Middle Cambrian. A double-punch mass extinction like this is reminiscent of both the Permo-Triassic and Early Jurassic crises, and there are other similarities too—black shales became widespread around this time and the carbon isotopes changed. However, explaining how all these Cambrian phenomenon fit together is still at an early and rather speculative stage.

The other proposed LIP–mass extinction link of the pre-Pangean world occurred around 375 million years ago at the boundary between the Frasnian and Famennian Stages of the Devonian. This F-F event, as it is known, was severe, and it has been linked with a little-known LIP in Siberia (not the Siberian Traps, but a smaller one further east), but once again the age dating is only suggestive, not conclusive.

We are still a long way from making a connection between volcanism and extinctions in the pre-Pangean world. Potential research avenues could involve looking for tell-tale clues such as trace metal concentrations in sedimentary rocks. Giant volcanism releases a lot of metals to the atmosphere and oceans, and this volcanism proxy is potentially preserved in rocks. The search has yet to be made for such evidence.

In summary, life since the Cambrian has gone through four broad stages: an initial 180-million-year-long interval prone to mass extinctions and lasting from the Cambrian to the Devonian. It was followed by a more pleasant 100-million-year interlude lacking mass extinctions (and LIPs), during the Carboniferous-Middle Permian interval. Then came Pangea and 80 million years of volcanic hell. Finally, Pangea broke up, and it has been 180 million years of smooth sailing ever since, despite numerous LIP eruptions and one terrible meteorite impact. The discussion of the first two stages requires another book, but I would briefly note that the first stage lacked terrestrial vegetation; the first forests did not appear until toward the end of the Devonian, as the final extinctions struck. There may be a link between these observations.

RESILIENT PLANET

The key point of this book has to been to ask, why, during the reign of Pangea, were LIPs and mass extinctions occurring in lockstep, whereas the connection has been broken since Pangea disintegrated? This lack of extinctions is despite the

observation that post-Pangea, LIPs still seem capable of causing interesting and significant global environmental change,
especially global warming and ocean deoxygenation. And
yet, none have been capable of causing major life crises. Why
have LIPs and coincidental environmental changes lost their
lethality?

To answer the question requires a return to the issue of
how the earth's ocean-climate system operates and specifically
how greenhouse gases are removed from the atmosphere. This
is the bread-and-butter of climate modelers such as Yannick
Donnadieu of the Laboratoire des Sciences du Climat et de
l'Environnement. Using the GEOCLIM modeling program,
he showed that the Pangean world was generally very hot and
arid with atmospheric CO_2 values typically around 3000 ppm.
By the Early Cretaceous, the world was much more efficient
at removing CO_2, with the result that atmospheric concentrations were maintained at an equilibrium value that was only
a tenth of the Pangean level. The reason is related to the scale
of Pangea. A huge continent has vast areas in the interior that
are too far away from the sea to receive much rain. In contrast, smaller, more fragmented continents receive precipitation over a greater area. The crucial point here is that rainfall
causes weathering, which consumes atmospheric CO_2 that is
then converted to the biocarbonate ion (HCO_3^-), which runs
off into the ocean, increasing ocean alkalinity and making
it easier to precipitate limestone. Continental runoff also
supplies nutrients to the oceans, which stimulate plankton
growth that removes more carbon dioxide, which gets buried
as organic carbon in marine sedimentary rocks.

There are two further important factors that make a supercontinent poor at removing atmospheric carbon dioxide.
Limestone deposition occurs in shallow shelf seas (because this
is where the animals live that secrete the carbonate skeletons
that form the limestones) and during Pangean times shelf area

was at a minimum. This can be explained by simple observation: the shelf fringe of one supercontinent is a lot less than the shelf length of several smaller continents of overall equal area. Consider the following example. A circle of area 4 square centimeters has a circumference of 7.1 centimeters, whereas four circles with an area of 1 square centimeter have a combined circumference of more than 15 centimeters. Pangea was equivalent to a big circle, so there was not that much shelf space for limestone to accumulate; this way of burying carbon was not very effective. Limited shelf-sea space also ensures that the other main mechanism for carbon removal—organic-matter burial in marine sediment—was not working well either because most organic material is buried in shelf sediments.

Finally, as we have seen, the rise of coccoliths further increased the efficiency of the carbon dioxide drawdown pump. Importantly, since Pangea broke apart, calcareous plankton limestone deposition is no longer just restricted to shallow seas; it also accumulates in deep-water settings. There is thus an enormous increase in the potential area available for limestones to form, and a much more rapid response to changes in ocean pH is ensured. Alkalinity increases in the past caused by, for example, increased global humidity and enhanced runoff of bicarbonate-laden river waters could not be readily balanced by increase limestone formation, because limestones formed only in the limited areas available in shelf seas, and their extent depended upon sea level and the length of continental perimeters. Consequently, an increased rain of planktonic carbonate into the deep seas can counterbalance episodes of elevated alkalinity while elevated acidity can be buffered by dissolving the huge amounts of carbonate sediments in the deep sea, thereby neutralizing the pH decline. Deep-sea carbonate is a superb buffer of changes in ocean pH.

There is another newly evolved group of plankton that has greatly increased the efficiency of carbon burial in sediments.

These are the diatoms, which appeared around the same time as the coccoliths, but only in the past few tens of millions of years have they become important. Diatoms are photosynthetic plankton that surround their organic cells with tiny skeletons of silica. This makes them much denser than other types of plankton that do not secrete a mineralized skeleton. Consequently, after their short life cycle in the surface waters, diatoms sink to the seabed quickly. This does not allow much time for their organic cells to decay in the water column, ensuring that more reach the sediment to be preserved and buried.

Terrestrial plants provide another important way of taking carbon dioxide out of the atmosphere (fig.7.1). Having been taken up by photosynthesis, the carbon can be buried in soil, stand around as the biomass of the plants, or return to the atmosphere as carbon dioxide from rotting vegetation. Around half of all the carbon buried in sediments today occurs on land, although the percentage varies enormously across geological time scales. But as a method of counterbalancing sudden increases in atmospheric CO_2 levels, carbon burial is not very effective because the rate on land cannot be changed quickly.

During episodes of increased availability of atmospheric carbon dioxide, plant productivity increases, because CO_2 is a valuable nutrient for plants, and forests become more luxuriant. So on a very short-term scale, CO_2 is removed from the atmosphere and converted to plant matter. This is why modern tropical rainforests, like the Amazon, are so important to the health of the world—they are mopping up a lot of the carbon we are putting into the atmosphere; however, this effect is only modest, and it is only temporary because it does not increase the amount of carbon buried in sediment. The soils on which the Amazon forest grows are, for example, very poor in organic matter. Most plant matter just decomposes back to carbon dioxide on the forest floor.

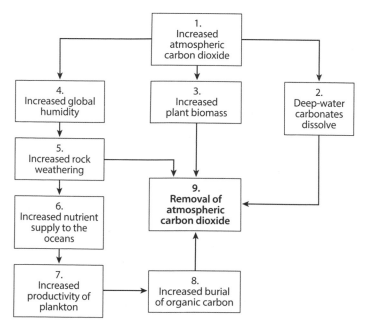

Figure 7.1. The carbon pump: how carbon dioxide is removed from the atmosphere by geological processes operating today.

To effectively bury terrestrial plant matter and so make an increased contribution to removal of atmospheric CO_2 requires peat accumulation, which occurs in low-lying, rapidly subsiding swamps (where coals form) in a humid climate. The availability of such settings is controlled by geological processes such as basin formation and so has varied greatly through geological time in a manner unrelated to climate fluctuations. The key issue here is that subsidence is not controlled by climate. This situation is directly analogous to the process of limestone formation before the coccoliths appeared, because geological processes (like supercontinent break-up) control the area of shelf seas available for limestone formation, and this rock formation is not affected by short-term fluctuations in factors like atmosphere CO_2 levels. Now that limestones also form in the deep sea, there is much less geological control on their formation but a much closer link to the climate.

Coal formation was extensive in the northern hemisphere during the Carboniferous Period, when low-lying swamps stretched across a vast distance from Pennsylvania through Europe to western Russia. In these regions, extensive equatorial basins proved to be the crucial factor for this coal-forming period, not the climate. In the Permian, such conditions were more prevalent in the high southern latitudes of Gondwana, and most coal formation was focused there at that time. In contrast, for much of the Cretaceous, climates were warm and humid—ideal for plant growth—and yet coal formation was less common at such times because most potential coal-forming basins were under water due to high sea levels. This illustrates the point that geology, not climate, controlled the amount of coal formation.

Burial of plant matter is therefore not intimately linked to the feedback mechanisms that govern the cycling of carbon between the atmosphere and the crust on short geological time scales. By this, I mean it is not very responsive to variations in atmospheric composition, although plants still play a role. Rock weathering is, for example, strongly controlled by organic acids generated by plant roots. Thus, increased humidity in a warmer world will favor plant growth, which in turn will favor accelerated rock weathering, which uses up atmospheric carbon dioxide. Even in the absence of plants, this weathering feedback process will still occur; it is just that plants speed it up. Thus, plants can be important and a world without plants would be much more prone to rapid climatic fluctuations. It was probably a significant factor following the mass extinction of plants at the end of the Permian too.

The efficiency and response time of carbon dioxide drawdown since the disintegration of Pangea has improved, with the result that sudden increases of atmospheric concentrations, such as seen at the onset of LIP eruptions, are no longer capable of causing catastrophes. There are now numerous ways in which the extra carbon in the atmosphere can be rapidly counterbalanced by extra carbon burial, and even in the short term, enhanced dissolution of deep-water carbonates is an effective buffer. Many of the improvements in the Earth's feedback mechanism relate to the break-up of Pangea, such as the higher overall humidity of smaller continents and the extra area available for carbon burial in the fringing shelf seas of dispersed continents. In addition, the subsequent evolution of planktonic groups with hard parts has greatly speeded up the transfer of carbon to sediments. These changes seem sufficient to stop the runaway climatic effects that cause mass extinctions. In some cases, such as the impressively large NAIP eruptions 60 million years ago, the environmental and climatic effects travel sufficiently far down the road that they begin to resemble a mass extinction but do not go far enough to actually cause one. The warming, acidification, and deep-ocean anoxia never get to the point where they become widespread and/or lethal.

During Pangea's existence, the effects of sudden climate warming became so severe that nasty feedback effects kicked in and magnified the consequences of warming to a degree that was catastrophic. At the greatest extreme, the failure to stem temperature rise during the Permo-Triassic crisis reached a point at which it became difficult to bury organic matter (because it decays quicker at higher temperatures). The result was temperatures too hot for terrestrial photosynthesis, which in turn retarded organic-matter formation and rock weathering, thereby weakening key parts of carbon cycle. Fortunately, the development of ocean anoxia meant that a new site of organic-matter burial appeared—the deep and cool abyssal depths—allowing normal conditions to be eventually reestablished. It was too late for most marine life, though, because by this time it had had to endure the double whammy of anoxia and high temperature.

The modern world's feedback mechanisms may be likened to a six-cylinder car engine that runs smoothly, occasionally speeding up when more CO_2 is injected but never misfiring (fig. 7.1). In contrast, Pangea's engine was four-cylinder and sometimes misfired and ran on three. And on at least one occasion, one process—terrestrial plant growth and burial—effectively shut down.

We still do not have all the answers to the questions posed by Pangean mass extinctions. The devastation of land communities is especially hard to explain. The extinction of Late Permian terrestrial communities is a truly awesome phenomenon, which might be related to atmospheric changes such as ozone destruction. Massive volcanic halogen emissions provide one cause, but then why did this only happen during Pangea's lifetime? More recent eruptions, such as those of the NAIP, would be expected to also emit huge amounts of halogens. Alternatively, some geologists have attempted to link changes in the oceans to events on land. Thus anoxic oceans may have leaked hydrogen sulfide into the atmosphere, where it would interfere with the formation of ozone. Unfortunately, this probably is not the answer. Hydrogen sulfide is immensely reactive with oxygen and is unlikely to ever reach the stratosphere and damage the ozone shield; instead, it will oxidize rapidly close to the sea surface. Perhaps terrestrial warming was the stress factor on land, but it would have been most harsh on forests adapted to living in cold conditions. The extinctions show no such temperature dependence—equatorial forests suffered as badly as those at high latitudes. And so the puzzle remains.

Clearly we do not yet have all the answers, but we have a come along way on the road to understanding the great mass extinctions of the fossil record.

This book is not intended as a warning from prehistory. If anything, its message has been the opposite—it was dangerous to live at a time of a supercontinent, whereas worlds like our own, with widely dispersed continents, are much better. They seem capable of absorbing the impact of sudden, dramatic increases of atmospheric carbon dioxide without any catastrophic change; surely an optimistic conclusion. Much of the "rapidity" of mass extinctions is in any case only rapid to a geologist—collapses in diversity typically happened in thousands of years—intervals that are almost beyond our concern. Would we worry if our actions today were storing up trouble for the world in 5000 or 10,000 years' time? Nonetheless, LIP eruptions are the only natural process to provide an analogue for our current impact on the atmosphere, and some of the more recent examples appear capable of causing pretty spectacular climate changes, such as the intense warming episode 60 million years ago. Even if the extinction losses at that time were very modest, the effects were significant; most life on land had to shift its distribution poleward as climatic belts expanded away from the equator. This was at a time before man-made landscape change had broken up natural habitats into fragmented, undisturbed oases in a sea of farmland and urban sprawl. Modern life on land is going to have a hard time moving (hopping between fragments) to accommodate warming trends.

In the oceans, the clear story from past mass extinctions is that global warming goes hand in hand with oxygen starvation and that both are very bad for life. The modern oceans are showing the first signs of these changes, and modeling predictions suggest that we may only be a few hundred years away from seeing large expanses of anoxic waters develop. The effect on sea life will not be good, although by then massive overfishing will probably already have destroyed most ecosystems as we currently know them.

The ultimate take-home message of this book is that for 80 million years a series of gigantic volcanic eruptions caused a series of crises that fundamentally changed the course of life and removed groups that would otherwise have thrived. We should pity the animals of Pangea. For the pareiasaurs, dinocephalians, and crurotarsans, all looked to be going so well, but they had their reigns cut unexpectedly short. In recompense it allowed our favorite animals, the dinosaurs, to take center stage for a very long time indeed, but they were simply lucky that their reign coincided with the break-up of a lethally large supercontinent. In the end their luck ran out too.

NOTES

1. If there were a prize for the most cumbersome and inconveniently long name ever devised for an interval of time, then this zone would be the clear winner. To make it a little more manageable I will just call it the *J. xuanhanensis* Zone.

2. We were not the only ones to identify the violent nature of Emeishan volcanism. Ingrid Ukstins-Peate of the University of Iowa and Scott Bryan of the Queensland University of Technology were also working in the region and recognized the significance of the volcanic breccias in a paper published in 2008.

3. Note that pH relates to the concentration of hydrogen ions in solution; values below 7 are acidic and values above 7 are alkali. When pH = 7, the water is said to be neutral. Today's ocean waters have are slightly alkaline (pH > 7), and this has probably been the case for much of geological history. The clearest evidence for this situation is the presence of limestones, which would dissolve were the oceans acidic.

4. Further evidence for this warming comes from Greg Retallack's analysis of fossil soil types in the region, which reveals the appearance of the mineral berthierene, an indicator of warm, humid conditions.

5. In fact, the first detailed study (in 1954) of the timing of the Permian extinction, by the German paleontologist Otto Schindewolf, concluded that it happened instantaneously and was caused by cosmic radiation from a supernova. Catastrophism was out of favor at the time, and his idea gained no adherents even though his basic observation—abrupt extinction—was correct.

6. This is known as a euxinic environment and was named after the Black Sea, which was called the Pontus Euxinus in Greek and Roman times.

7. Note that the ratio $^{13}C/^{12}C$ is usually compared relative to the ratio of a standard and is expressed as a $\delta 13C$ value. Thus, a sample with a value of 0 ‰ indicates it has the same ratio as the standard while a value of +4 ‰ is a sample that is 4 parts per thousand heavier. The symbol ‰ stands for "per mil," or "per thousand."

8. "Primitive" is used here in the sense that it is original mantle, which was present when the Earth first cooled. It has not subsequently been involved in magma generation. Virgin mantle would be an alternative name.

9. This result has been called the Lilliput effect and is discussed further in chapter 4.

10. Robert Toggweiler and Joellen Russell (2008) provided a useful state-of-the-art perspective of our rapidly changing understanding of the ocean-climate system that shows how fast ideas are changing. If there is still much to learn about the workings of the modern world, then pity those trying to reconstruct conditions more than 250 million years ago!

11. Worms actually have virtually no fossilization potential, but their burrows are diverse and common, and we know from these that many burrow makers disappeared during the Permo-Triassic mass extinction.

12. This 2014 paper by Clara Mackenzie and her Bangor colleagues is listed in the references along with the other key papers on this highly debated topic.

13. In fact, generic losses probably exceeded 50%, which is of course very high, but by the standards of the Permo-Triassic mass extinction, it was not too bad.

14. Recent zircon dating by M. T. Paton (University of Western Australia) of Siberian Traps sills suggests that there was a phase of volcanism, younger than the main phase—around 249.6 million years ago. This could, just possibly, indicate a phase of eruption at the S/S boundary, but more such studies are needed.

15. It would be nice to know the values for the Early Triassic supergreenhouse too, but no equivalent study has been done because of the rarity of fossil leaves at this time, probably because much of the world was too hot to support vegetation.

16. The carbon isotope ratio in organic carbon and carbonate in soil is related, in a rather complicated way, to the atmospheric CO_2 concentration.

17. Jet is a variety of fossil driftwood much sought after because it can be carved into jewelry. It is not uncommon in the eponymous Jet Rock and was for many years mined in the cliffs around Whitby.

18. An English idiom derived from cricket that refers to the misfortune of a bowler.

19. In our 1997 book *Mass Extinctions and Their Aftermath*, Tony Hallam and I called this a minor mass extinction, which in hindsight sounds a bit oxymoronic.

20. K. McAuliffe. 2008. Ocean reflux. *Discover* July 2008, 28–37.

REFERENCES

CHAPTER 1

Alvarez, L., W. Alvarez, F. Asaro, and H. V. Michel. 1980. Extraterrestrial cause for the Cretaceous-Tertiary extinction—Experimental results and theoretical implications. *Science* 208: 1095–1108.

Benton, M. J. 2003. *When Life Nearly Died: The Greatest Mass Extinction of All Time*. London: Thames & Hudson.

Courtillot, V. 1999. *Evolutionary Catastrophes: The Science of Mass Extinction*. Cambridge: Cambridge University Press.

Hallam, A. 2004. *Catastrophes and Lesser Calamities: The Causes of Mass Extinctions*. Oxford: Oxford University Press.

Hallam, A., and P. B. Wignall. 1997. *Mass Extinctions and Their Aftermath*. Oxford: Oxford University Press.

Torsvik, T. H., and L.R.M Cocks. 2004. Earth geography from 400 to 250Ma: A palaeomagnetic, faunal and facies review. *J. Geol. Soc. London* 161: 555–572.

Wignall, P. B. 2001. Large igneous provinces and mass extinctions. *Earth-Science Rev.* 53: 1–33.

———. 2005. The link between large igneous province eruptions and mass extinctions. *Elements* 1: 293–297.

CHAPTER 2

Arche, A., and J. López-Gómez. 2005. Sudden change in fluvial style across the Permian-Triassic boundary in the eastern Iberian Ranges, Spain: Analysis of possible causes. *Palaeogeogr. Palaeoclimatol. Palaeoecol.* 229: 104–126.

Bambach, R. K., A. H. Knoll, and S. C. Wang. 2004. Origination, extinction, and mass depletions of marine diversity. *Paleobiology* 30: 522–542.

Bond, D.P.G., J. Hilton, P. B. Wignall, J. R. Ali, L. G. Stevens, Y.-D. Sun, and X.-L. Lai. 2010. The Middle Permian (Capitanian) mass extinction on land and in the oceans. *Earth-Science Rev.* 109: 100–116.

Chung, S-L., and B.-M. Jan. 1995. Plume-lithosphere interaction in generation of the Emeishan flood basalts at the Permian-Triassic boundary. *Geology* 23: 883–892.

Clapham, M.E., and J. L. Payne. 2011. Acidification, anoxia, and extinction: A multiple logistic regression analysis of extinction selectivity during the Middle and Late Permian. *Geology* 39: 1059–1062.

Courtillot, V. 1999. *Evolutionary Catastrophes: The Science of Mass Extinction.* Cambridge: Cambridge University Press.

Ganino, C., and N. T. Arndt. 2008. Climate changes caused by degassing of sediments during the emplacement of large igneous provinces. *Geology* 37: 323–326.

Isozaki, Y., and A. Ota. 2001. Middle-Upper Permian (Maokouan-Wuchiapingian) boundary in mid-oceanic paleo-atoll limestone of Kamura and Akasaka, Japan. *Proc. Japan Acad.* 77, Ser. B: 104–109.

Jin, Y.-G., J. Zhang, and Q.-H Shang. 1994. Two phases of the end-Permian mass extinction. In A. F. Embry, B. Beauchamp, and D. J. Glass (eds.), *Pangea: Global Environments and Resources*, 813–822. Calgary, Canada: Canadian Soc. Pet. Geol.

McGhee, Jr., G. R., M. E Clapham, P. M. Sheehan, D. J. Bottjer, and M. L. Droser. 2013. A new ecological-severity ranking of major Phanerozoic diversity crises. *Palaeogeogr. Palaeoclimatol. Palaeoecol.* 370: 260–270.

Retallack, G. J., C. A. Metzger, T. Greaver, A. H. Jahren, R.M.H. Smith, and N. D. Sheldon. 2006. Middle-Late Permian mass extinction on land. *Bull. Geol. Soc. Amer.* 118: 1398–1411.

Shen, S.-Z., and G. R. Shi. 1996. Diversity and extinction patterns of Permian Brachiopoda of South China. *Hist. Biol.* 12: 93–110.

Stanley, S. M., and X.-N Yang. 1994. A double mass extinction at the end of the Paleozoic Era. *Science* 266: 1340–1344.

Svensen, H., S. Planke, A. Malthe-Sørenssen, B. Jamtveit, R. Myklebust, T. Rasmussen Eldem, and S. S. Rey. 2004. Release of methane from a volcanic basin as a mechanism for initial Eocene global warming. *Nature* 429: 542–545.

Ukstins Peate, I., and S. E. Bryan. 2008. Re-evaluating plume-induced uplift in the Emeishan large igneous province. *Nature Geosci.* 1: 625–629.

Wignall, P. B., Y.-D. Sun, D.P.G. Bond, G. Izon, R. J. Newton, S. Védrine, M. Widdowson, J. R. Ali, X.-L. Lai, H.-S. Jiang, H. Cope, and S. H. Bottrell. 2009. Volcanism, mass extinction and carbon isotope fluctuations in the Middle Permian of China. *Science* 324: 1179–1182.

CHAPTER 3

Andersson, A. J., F. T. Mackenzie, and J.-P. Gattuso. 2011. Effects of ocean acidification on benthic processes, organisms, and ecosystems. In J. P. Gattuso, and L. Hansson (eds.), *Ocean Acidification*, 122–153. Oxford: Oxford University Press.

Beerling, D. J., M. Harfoot, B. Lomax, and J. A. Pyle. 2007. The stability of the stratospheric ozone layer during the end-Permian eruption of the Siberian Traps. *Phil. Trans. Royal Soc. A* 365: 1843–1866.

Benton, M. J., and A. J. Newell. 2014. Impacts of global warming on Permo-Triassic terrestrial ecosystems. *Gondwana Res.* 25: 1308–1337.

Bond, D.P.G., and P. B. Wignall. 2010. Pyrite framboid study of marine Permo-Triassic boundary sections: A complex anoxic event and its relationship to contemporaneous mass extinction. *Bull. Geol. Soc. Amer.* 122: 1265–1279.

Campbell, I. H., G. K. Czamanske, V. A. Fedorenko, R. I. Hill, and V. Stepanov. 1992. Synchronism of the Siberian Traps and Permian-Triassic boundary. *Science* 258: 1760–1763.

Ehrenberg, S. N., T. A. Svånå, and P. K. Swart. 2008. Uranium depletion across the Permian-Triassic boundary in Middle East carbonates: Signature of oceanic anoxia. *Bull. Amer. Assoc. Petrol. Geol.* 92: 691–707.

Erwin, D. H. 1993. *The Great Paleozoic Crisis: Life and Death in the Permian.* New York: Columbia University Press.

Friedman, M., and L. C. Sallan. 2012. Five hundred million years of extinction and recovery: A Phanerozoic survey of large-scale diversity patterns in fishes. *Palaeontology* 55: 707–742.

Grasby, S. E., H. Sanei, and B. Beauchamp. 2011. Catastrophic dispersion of coal fly ash into oceans during the latest Permian extinction. *Nature Geosci.* 4: 104–107.

Grice, K., C.-Q. Cao, G. D. Love, M. E. Böttcher, R. J. Twitchett, E. Grosjean, R. E. Summons, S. C. Turgeon, W. Dunning, and Y.-G Jin. 2005. Photic zone euxinia during the Permian-Triassic superanoxic event. *Science* 307: 706–709.

Heydari, E., and J. Hassanzadeh. 2003. Deev Jahi Model of the Permian-Triassic boundary mass extinction: A case for gas hydrates as the main cause of biological crisis on Earth. *Sedi. Geol.* 163: 147–163.

Hori, R. S., S. Yamakita, M. Ikehara, K. Kodama, Y. Aita, T. Sakai, A. Takemura, Y. Kamata, N. Suzuki, S. Takahashi, K. B. Spörli, and J. A. Grant-Mackie. 2011. Early Triassic (Induan) Radiolaria and carbon-isotope ratios of deep-sea sequence from Waiheke Island, North Island, New Zealand. *Palaeoworld* 20: 166–178.

Isozaki, Y. 1997. Permo-Triassic boundary superanoxia and stratified superocean: Records from lost deep sea. *Science* 276: 235–238.

Joachimski, M. M., X.-L. Lai, S.-Z. Shen, H.-S. Jiang, G.-M. Luo, B. Chen, J. Chen, and Y.-D. Sun. 2012. Climate warming in the latest Permian and Permian-Triassic mass extinction. *Geology* 40: 195–198.

Kaiho, K., Y. Kajiwara, Z.-Q. Chen, and P. Gorjan. 2006. A sulfur isotope event at the end of the Permian. *Chem. Geol.* 235, 33–47.

Knoll, A. H., R. K. Bambach, J. L. Payne, S. Pruss, and W. W. Fischer. 2007. Paleophysiology and end-Permian mass extinction. *Earth Planet. Sci. Letts.* 256: 295–313.

Kozur, H. W., and R. E. Weems. 2011. Detailed correlation and age of continental late Changhsingian and earliest Triassic beds: Implications for the role of the Siberian Trap in the Permian-Triassic biotic crisis. *Palaeogeogr. Palaeoclimatol. Palaeoecol.* 308: 22–40.

Lindström, S., and S. McLoughlin. 2007. Synchronous palynofloristic extinction and recovery after the end-Permian event in the Prince Charles Mountains, Antarctica: Implications for palynofloristic turnover across Gondwana. *Rev. Palaeobotany Palynology* 145: 89–122.

Looy, C. V., R. J. Twitchett, D. L. Dilcher, J.H.A. Van Konijnenburg-Van Cittert, and H. Visscher. 2001. Life in the end-Permian dead zone. *Proc. Natl. Acad. Sci.* 98: 7879–7883.

Mackenzie, C. L., G. A. Ormondroyd, S. F. Curling, R. J. Ball, N. M. Whiteley, and S. K. Malham. 2014. Ocean warming, more than acidification, reduces shell strength in a commercial shellfish species during food limitation. *PLOS One* 9, e86764. doi:10.1371/journal.pone.0086764.

Meyer, K. M., L R. Kump, and A. Ridgwell. 2008. Biogeochemical controls on photic-zone euxinia during the end-Permian mass extinction. *Geology* 42: 747–750.

Montenegro, A., P. Spence, K. J. Meissner, M. Eby, M. J. Melchin, and S. T. Johnston. 2011. Climate simulations of the Permian-Triassic boundary: Ocean acidification and the acidification event. *Paleoceanography* 26. doi:10.1029/2010PA002058.

Payne, J. L., D. J. Lehrmann, D. Follett, M. Seibel, L. R. Kump, A. Riccardi, D. Altiner, H. Sano, and J.-Y. Wei. 2007. Erosional truncation of uppermost Permian shallow-marine carbonates and implications for Permian-Triassic boundary events. *Bull. Geol. Soc. Amer.* 119: 771–784.

Payne, J. L., D. J. Lehrmann, J.-Y. Wei, M. J. Orchard, D. P. Schrag, and A. H. Knoll. 2004. Large perturbations of the carbon cycle during recovery from the end-Permian extinction. *Science* 305: 506–509.

Peng, Y.-Q., and G. R. Shi. 2009. Life crises on land across the Permian-Triassic boundary in South China. *Global Planet. Change* 65: 155–165.

Pörtner, H. O. 2010 Oxygen- and capacity-limitation of thermal tolerance: A matrix for integrating climate-related stressor effects in marine ecosystems. *J. Exp. Biol.* 213: 881–893.

Reichow, M. K., M. S. Pringle, A. I. Al'Mukhamedov, M. B. Allen, V. L. Andreichev, M. M. Buslov, C. E. Davies, G. S. Fedoseev, J. G. Fitton, S. Inger, A. Ya. Medvedev, C. Mitchell, V. N. Puchkov, I. Yu. Safanova, R. A. Scott, and A. D. Saunders. 2009. The timing and extent of the eruption of the Siberian Traps large igneous province: Implications for the end-Permian environmental crisis. *Earth Planet. Sci. Letts.* 277: 9–20.

Scherbakov, D. E. 2008. Insect recovery after the Permian/Triassic crisis. *Alavesia* 2: 125–131.

Schindewolf, O. H. 1954 Über die möglichen Ursachen der grossen erdgeschtlichen Faunenschnitte. *Neues Jahrbuch Geol. Paläontol. Monat.* 1954: 457–465.

Sephton, M. A., C. V. Looy, H. Brinkhuis, P. B. Wignall, J. W. de Leeuw, and H. Visscher. 2005. Catastrophic soil erosion during the end-Permian biotic crisis. *Geology* 33: 941–44.

Shen, S. Z., J. L. Crowley, Y. Wang, S. A. Bowring, D. H. Erwin, P. M. Sadler, C. Q. Cao, D. H. Rothman, C. M. Henderson, J. Ramezani, H. Zhang, Y. Shen, X. D. Wang, W. Wang, L. Mu, W. Z. Li, Y. G. Tang, X. L. Liu, L. J. Liu, Y. Zeng, Y. F. Jiang, and Y. G. Jin. 2011. Calibrating the end-Permian mass extinction. *Science* 334: 1367–72.

Sobolev, S. V., A. V. Sobolev, D. V. Kuzmin, N. A. Krivolutskaya, A. G. Petrunin, N. T. Arndt, V. A. Radko, and Y. R. Vasiliev. 2011. Linking mantle plumes, large igneous provinces and environmental catastrophes. *Nature* 477: 312–316.

Song, H.J., P. B. Wignall, D. L. Chu, J. N. Tong, Y. D. Sun, H. Y. Song, W. H. He, and T. Li. 2014. Anoxia/high temperature double whammy during the Permian-Triassic marine crisis and its aftermath. *Sci. Rep.* 4: 4132. doi:10.1038/srep04132.

Song, H.-J., P. B. Wignall, J. N. Tong, and H.-F. Yin. 2013. Two pulses of extinction during the Permian-Triassic crisis. *Nature Geosci.* 6: 52–56.

Toggweiler, J. R., and J. Russell. 2008. Ocean circulation in a warming climate. *Nature* 451: 286–288.

Visscher, H., H. Brinkhuis, D. L. Dilcher, W. C. Elsik, Y. Eshet, c. V. Looy, M. R. Rampino, and A. Traverse. 1996. The terminal Paleozoic fungal event: Evidence of terrestrial ecosystem destabilisation. *Proc. Natl. Acad. Sci.* 93: 2155–2158.

Visscher, H., C. V. Looy, M. E. Collinson, H. Brinkhuis, J.H.A. van Konijnenburg-Cittert, W. M. Kürschner, and M. A. Sephton. 2004. Environmental mutagenesis during the end-Permian ecological crisis. *Proc. Natl. Acad. Sci.* 101: 12952–12956.

Ward, P. D., D. R. Montgomery, and R. Smith, R. 2000. Altered river morphology in South Africa related to the Permian-Triassic extinction. *Science* 289: 1740–1743.

Wignall, P. B., D.P.G. Bond, K. Kuwahara, K. Kakuwa, R. J. Newton, and S. W. Poulton. 2010. An 80 million year oceanic redox history from Permian to Jurassic pelagic sediments of the Mino-Tamba terrane, SW Japan, and the origin of four mass extinctions. *Global Planet. Change* 71: 109–123.

Wignall, P. B., and A. Hallam. 1992. Anoxia as a cause of the Permian/Triassic extinction: Facies evidence from northern Italy and the western United States. *Palaeogeogr. Palaeoclimatol. Palaeoecol.* 93: 21–46.

Wignall, P. B., S. Kershaw, P.-Y. Collin, and S. Crasquin-Soleau. 2009. Erosional truncation of uppermost Permian shallow-marine carbonates and implications for Permian-Triassic boundary events: Comment. *Bull. Geol. Soc. Amer.* 121: 954–956.

Wignall, P. B., and R. Newton. 2003. Contrasting deep-water records from the Upper Permian and Lower Triassic of South Tibet and British Columbia: Evidence for a diachronous mass extinction. *Palaios* 18: 153–167.

Winguth, A.M.E., and E. Maier-Reimer. 2005. Causes of marine productivity and oxygen changes associated with the Permian-Triassic boundary: A reevaluation with ocean general circulation models. *Mar. Geol.* 217: 283–304.

Winguth, C., and A.M.E. Winguth. 2012 Simulating Permian-Triassic oceanic anoxia distribution: Implications for species extinction and recovery. *Geology* 40: 127–130.

Wittmann, A. C., and H. O. Pörtner. 2013. Sensitivies of extant animal taxa to ocean acidification. *Nature Clim. Change* 3: 995–1001.

CHAPTER 4

Benton, M. J. 1986. More than one event in the late Triassic mass extinction. *Nature* 321: 857–861.

Brayard, A., G. Escarguel, H. Bucher, C. Monnet, T. Brühwiller, N. Goudemand, T. Galfetti, and J. Guex. 2009. Good genes and good luck: Ammonoids diversity and the end-Permian mass extinction. *Science* 325: 1118–1121.

Brusatte, S. L., S. J. Nesbitt, R. B. Irmis, R. J. Butler, M. J. Benton, and M. A. Norell. 2010. The origin and early radiation of the dinosaurs. *Earth-Science Rev.* 101: 68–100.

Dal Coso, J., P. Mietto, R. J. Newton, R. D. Pancost, N. Preto, G. Roghi, and P. B. Wignall. 2012. Discover of a major negative $\delta^{13}C$ spike in the Carnian (Late Triassic) linked to eruption of Wrangelia flood basalts. *Geology* 40: 79–82.

Falkowski, P. G., M. E. Katz, A. H. Knoll, A. Quigg, J. A. Raven, O. Schofield, and F. J. Taylor. 2004. The evolution of modern eukaryotic phytoplankton. *Science* 305: 352–360.

Fröbisch, N. B., J. Fröbisch, P. M. Sander, L. Schmitz, and O. Rieppl. 2013. Macropredatory ichthyosaur from the Middle Triassic and the origin of modern trophic networks. *Proc. Natl. Acad. Sci.* 110: 1393–1397.

Galfetti, T., P. A. Hochuli, A. Brayard, H. Bucher, H. Weissert, and J. Os Vigran. 2007. Smithian-Spathian boundary event: Evidence for global climatic change in the wake of the end-Permian biotic crisis. *Geology* 35: 291–294.

Goddéris, Y., Y. Donnadieu, C. de Vargas, R. T. Pierrehumbert, G. Dromart, and B. van de Schootbrugge. 2008 Causal or casual link between the rise of nannoplankton calcification and a tectonically-driven massive decrease in Late Triassic atmospheric CO_2? *Earth Planet. Sci. Letts.* 267: 247–255.

Kaim, A., and A. Nützel. 2011. Dead bellerophontids walking—The short Mesozoic history of the Bellerophontoidea (Gastropoda). *Palaeogeogr. Palaeoclimatol. Palaeoecol.* 308: 190–199.

McGowan, A. J. 2005 Ammonoid recovery from the Late Permian mass extinction event. *Compt. Rendus Palevol.* 4: 449–462.

Nakada, R., K. Ogawa, N. Suzuki, S. Takahashi, and Y. Takahashi. 2014. Late Triassic compositional changes of aeolian dusts in the pelagic Panthalassa: Response to the continental climatic change. *Palaeogeogr. Palaeoclimatol. Palaeoecol.* 393: 61–75.

Onue, T., and A. Yoshida. 2010. Depositional response to the Late Triassic ascent of calcareous plankton in pelagic mid-oceanic plate deposits of Japan. *J. Asian Ear. Sci.* 37: 312–321.

Paton, M. T., A. V. Ivanov, M. L. Fiorentini, N. J. McNaughton, I. Mudrovska, L. Z. Reznitski, and E. I. Dementerova. 2010. Late Permian and Early Triassic magmatic pulses in the Angara-Taseeva syncline, southern Siberian Traps and their possible influence on the environment. *Russian Geol. Geophys.* 51: 1012–1020.

Payne, J. L., and L. R. Kump. 2007. Evidence for recurrent Early Triassic massive volcanism from quantitative interpretation of carbon isotope fluctuations. *Earth Planet. Sci. Letts.* 256: 264–277.

Retallack, G. J. 1999. Postapocalyptic greenhouse paleoclimate revealed by earliest Triassic paleosols in the Sydney Basin, Australia. *Bull. Geol. Soc. Amer.* 111: 52–70.

Saito, R., K. Kaiho, M. Oba, S. Takahashi, Z.-Q. Chen, , and J.-N. Tong. 2013. A terrestrial vegetation turnover in the middle of the Early Triassic. *Global Planet. Change* 105: 152–159.

Simms, M. J., and A. H. Ruffell. 1989. Synchroneity of climatic change and extinctions in the Late Triassic. *Geology* 17: 265–268.

Song, H.-J., P. B. Wignall, Z.-Q. Chen, J.-N. Tong, D.P.G. Bond, X.-L. Lai, X.-M. Zhao, H.-S. Jiang, C.-B. Yan, Z.-J. Nin, J. Chen, H. Yang, and Y.-B. Wang. 2011. Recovery tempo and pattern of marine ecosystems after the end-Permian mass extinction. *Geology* 39: 739–742.

Sun, Y.-D., M. M. Joachimski, P. B. Wignall, C.-B. Yan, Y.-L. Chen, H.-S. Jiang, L.-N. Wang, and X.-L. Lai. 2012. Lethally hot temperatures during the Early Triassic greenhouse. *Science* 388: 366–370.

Twitchett, R. J., and P. B. Wignall. 1996. Trace fossils and the aftermath of the Permo-Triassic mass extinction: Evidence from northern Italy. *Palaeogeogr. Palaeoclimatol. Palaeoecol.* 124: 137–151.

CHAPTER 5

Ashraf, A. R., Y.-W. Sun, G. Sun, D. Uhl, V. Mosbruger, J. Li, and M. Herrmann. 2010. Triassic and Jurassic paleoclimate development in the Junggar Basin, Xinjiang, Northwest China—A review and additional lithological data. *Palaeobio. Palaeoenv.* 90: 187–201.

Benton, M. J. 1986. More than one event in the late Triassic mass extinction. *Nature* 321: 857–861.

Carter, E. S., and R. Hori. 2005. Global correlation of the radiolarian faunal change across the Triassic-Jurassic boundary. *Canadian J. Earth Sci.* 42: 777–790.

Hallam, A. 1981. The end-Triassic bivalve extinction event. *Palaeogeogr. Palaeoclimatol. Palaeoecol.* 35: 1–44.

———. 2004. *Catastrophes and Lesser Calamities.* Oxford: Oxford University Press.

Hallam, A., P. B. Wignall, J.-R. Yin, and R. Riding. 2000. An investigation into possible facies changes across the Triassic-Jurassic boundary in southern Tibet. *Sediment. Geol.* 137: 101–106.

Hautmann, M. 2004. Effect of end-Triassic CO_2 maximum on carbonate sedimentation and marine mass-extinction. *Facies* 53: 257–261.

Hesselbo, S. P., S. A. Robinson, F. Surlyk, and S. Piasecki. 2002. Terrestrial and marine extinction at the Triassic-Jurassic boundary synchronized with major carbon cycle perturbation: A link to initiation of massive volcanism. *Geology* 30: 251–54.

Hori, R. S., T. Fujiki, E. Inoue, and J.-I. Kimura. 2007. Platinum group element anomalies and bioevents in Triassic-Jurassic deep-sea sediments of Panthalassa *Paleogeog. Paleoclim. Palaeoecol.* 244: 391–406.

Kuroda, J., R. S. Hori, K. Suzuki, D. R. Gröcke, and N. Ohkuchi. 2010. Marine osmium isotope record across the Triassic-Jurassic boundary from a Pacific pelagic site. *Geology* 38: 1095–1098.

Marzoli, A., P. R. Renne, E. M. Piccirillo, A. Ernesto, G. Bellieni, and A. De Min. 1999. Extensive 200-million-year-old continental flood basalts of the Central Atlantic Magmatic Province. *Science* 284: 616–618.

McElwain, J. C., D. J. Beerling, and F. I. Woodward. 1999. Fossil plants and global warming at the Triassic-Jurassic boundary. *Science* 285: 1386–1390.

Richoz, S., B. van de Schootbrugge, J. Pross, W. Püttman, T. M. Quan, S. Lindström, C. Heunisch, J. Fiebig, R. Maquil, S. Schouten, C. A. Hauzenberger, and P. B. Wignall. 2012. Hydrogen sulphide poisoning of shallow seas following the end-Triassic extinction. *Nature Geosci.* 5: 662–667.

Schaller, M. F., J. D. Wright, and D. V. Kent. 2011. Atmospheric pCO_2 perturbations associated with the Central Atlantic Magmatic Province. *Science* 331: 1404–1409.

Simms, M. J. 2007. Uniquely extensive soft-sediment deformation in the Rhaetian of the UK: Evidence for earthquake or impact? *Paleogeog. Paleoclim. Palaeoecol.* 244: 407–423.

Smith, P. L., L. M. Longridge, M. Grey, J. Zhang, and B. Liang. 2014. From near extinction to recovery: Late Triassic to Middle Jurassic ammonoid shell geometry. *Lethaia* 47: 337–351.

Tanner, L. H., S. G. Lucas, and M. G. Chapman. 2004. Assessing the record and causes of Late Triassic extinctions. *Earth-Science Rev.* 65: 103–139.

Thorne, P. M., M. Ruta, and M. J. Benton. 2011. Resetting the evolution of marine reptiles at the Triassic/Jurassic boundary. *Proc. Natl. Acad. Sci.* 108: 8339–8244.

van den Berg, T., D. I. Whiteside, P. Viegas, R. Schouten, and M. J. Benton. 2012. The Late Triassic microvertebrate fauna of Tytherington, UK. *Proc. Geol. Assoc.* 123: 638–648.

van de Schootbrugge, B., T. M. Quan, S. Lindström, W. Püttmann, C. Heunisch, J. Pross, J. Fiebig, R. Petchick, H.-G. Röhling, S. Richoz, Y. Rosenthal, and P. G. Falkowski. 2009. Floral change across the Triassic/Jurassic boundary linked to flood basalt volcanism. *Nature Geosci.* 2: 589–594.

Wignall, P. B., and D.P.G. Bond. 2008. The end-Triassic and Early Jurassic mass extinction records in the British Isles. *Proc. Geol. Assoc.* 119: 73–84.

CHAPTER 6

Bailey, T. R., Y. Rosenthal, J. M. McArthur, B. van de Schootbrugge, and M. F. Thirlwall. 2003. Paleoceanographic changes of the Late Pliensbachian-Early Toarcian interval: A possible link to the genesis of an Oceanic Anoxic Event. *Earth Planet. Sci. Letts.* 212: 307–320.

Carrano, M. T., R.B.J. Benson, and S. D. Sampson. 2012. The phylogeny of Tetanura (Dinosauria: Theropoda). *J. System. Palaeo.* 10: 211–300.

Gómez, J. J., and A. Goy, A. 2011. Warming-driven mass extinction in the Early Toarcian (Early Jurassic) of northern and central Spain. Correlation with other time-equivalent European sections. *Palaeogeogr. Palaeoclimatol. Palaeoecol.* 306: 176–195.

Goričan, S., E. S. Carter, J. Guex, L. O'Dogherty, P. De Wever, P. Dumitrica, R. S. Hori, A. Matsuoka, and P. A. Whalen. 2013. Evolutionary patterns and palaeobiogeography of Pliensbachian and Toarcian (Early Jurassic) radiolarian. *Palaeogeogr. Palaeoclimatol. Palaeoecol.* 386: 620–636.

Hallam, A. 1961. Cyclothems, transgressions and faunal change in the Lias of north-west Europe. *Trans. Edinburgh Geol. Soc.* 18: 124–174.

Little, C.T.S., and M. J. Benton. 1995. Early Jurassic mass extinction: A global long-term event. *Geology* 23: 495–498.

Mattioli, E., B. Pittet, L. Petitpierre, and S. Maillot. 2009. Dramatic decrease of pelagic carbonate production by nannoplankton across the Early Toarcian anoxic event (T-OAE). *Global Planet. Change* 65: 134–145.

McElwain, J., J. Wade-Murphy, and S. P. Hesselbo. 2005. Changes in carbon dioxide during an oceanic anoxic event linked to intrusion into Gondwana coals. *Nature* 435: 479–482.

Pálfy, J., and P. L. Smith. 2000. Synchrony between Early Jurassic extinction, oceanic anoxic event, and the Karoo-Ferrar flood basalt volcanism. *Geology* 28: 747–750.

Sell, B., M. Ovtcharova, J. Guex, A. Bartolini, F. Jourdan, J. E. Spangenberg, J.-C. Vicente, and U. Schaltegger. 2014. Evaluating the temporal link between Karoo LIP and climatic-biologic events of the Toarcian Stage with high-precision U-Pb geochronology. *Earth Planet. Sci. Letts.* 408: 48–56.

Svensen, H., S. Planke, L. Chevallier, A. Malthe-Sørenssen, F. Corfu, and B. Jamtveit. 2007. Hydrothermal venting of greenhouse gases triggering Early Jurassic global warming. *Earth Planet. Sci. Letts.* 256: 554–566.

Thorne, P. M., M. Ruta, and M. J. Benton. 2011. Resetting the evolution of marine reptiles at the Triassic-Jurassic boundary. *Proc. Natl. Acad. Sci.* 108: 8339–8344.

Wignall, P. B., A. Hallam, R. J. Newton, J.-G. Sha, E. Reeves, E. Mattioli, and S. Crowley. 2006. An eastern Tethyan (Tibetan) record of the Early Jurassic (Toarcian) mass extinction event. *Geobiology* 4: 179–190.

Wignall, P. B., R. A. Newton, and C.T.S. Little. 2005. The timing of paleo-environmental change and cause-and-effect relationships during the Early Jurassic mass extinction in Europe. *American J. Sci.* 305: 1014–1032.

CHAPTER 7

Brusatte, S. L., R. J. Butler, P. M. Barrett, M. T. Carrano, D. C. Evans, G. T. Lloyd, P. D. Mannion, M. A. Norell, D. J. Peppe, P. Upchurch, and T. E. Williamson. 2014. The extinction of the dinosaurs. *Biol. Rev.* doi:10.1111/brv.12128.

Callegaro, S., D. R. Baker, A. De Min, A. Marzoli, K. Geraki, H. Bertrand, C. Viti, and F. Nestola. 2014. Microanalyses link sulfur from large igneous provinces and Mesozoic mass extinctions. *Geology* 42: 895–898.

Courtillot, V., V. A. Kravchinsky, X. Quidelleur, P. R. Renne, and D. P. Gladkochub. 2010. Preliminary dating of the Viluy traps (Eastern Siberia): Eruption at the time of Late Devonian extinction events? *Earth Planet. Sci. Letts.* 300: 239–245.

Donnadieu, Y., Y. Goddéris, R. Pierrehumbert, G. Dromart, F. Fluteau, and R. Jacob. 2006. A GEOCLIM simulation of climatic and biogeochemical consequences of Pangea breakup. *Geochem. Geophys. Geosystems* 7: Q11019. doi:10.1029/2006GC001278.

Föllmi, K. B. 2012. Early Cretaceous life, climate and anoxia. *Cretaceous Res.* 35: 230–257.

Gibbs, S. M., H. M. Stoll, P. R. Bown, and T. J. Bralower. 2010. Ocean acidification and surface water carbonate production across the Paleocene-Eocene thermal maximum. *Earth Planet. Sci. Letts.* 295: 583–592.

Gingerich, P. D. 2006. Environment and evolution through the Paleocene-Eocene thermal maximum. *Trends Ecol. Evol.* 21: 246–253.

Hallam, A., and P. B. Wignall. 1997. *Mass Extinctions and Their Aftermath.* Oxford: Oxford University Press.

Jourdan, F., K. Hodges, B. Sell, U. Schaltegger, M.T.D. Wingate, L. Z. Evins, U. Söderlund, P. W. Haines, D. Phillips, and T. Blenkinsop. 2014. High-precision dating of the Kalkarindji large igneous province, Australia, and synchrony with the Early-Middle Cambrian (Stage 4–5) extinction. *Geology* 42: 543–546.

Kump, L. R., T. J. Bralower, and A. Ridgwell. 2009. Ocean acidification in deep time. *Oceanography* 22: 94–107.

Svensen, H., S. Planke, A. Malthe-Sørenssen, B. Jamtveit, R. Myklebust, T. Rasmussen Eldem, and S. S. Rey. 2004. Release of methane from a volcanic basin as a mechanism for initial Eocene global warming. *Nature* 429: 542–545.

Wignall, P. B. 2001. Large igneous provinces and mass extinctions. *Earth-Science Rev.* 53: 1–33.

Wilfe, P., K. R. Johnson, and B. T. Huber. 2003. Correlated terrestrial and marine climate evidence for global changes before mass extinction at the Cretaceous-Paleogene boundary. *Proc. Natl. Acad. Sci.* 100: 599–604.

INDEX

acidification: Capitanian extinction, 36–38; ocean, 82–85, 83, 104, 116, 134, 152, 163–65, 173
actinopterygians, 14
Adelobasileus, 114
aetosaurs, 121
Ali, Jason, xii, 19, 22
Alpine limestones, 127–28
ammonites, 139–41, *plate 15*
ammonoids: Capitanian crisis, 14–16, 40; Jurassic Period, 96, 137; marine salinity and, 125; Permo-Triassic extinction, 52, 86, 93, 113; Rhaetian Stage, 130, 132; Smithian/Spathian extinction, 102; Triassic Period, 92–93, 105
amphibians, 2, 44–45, 95, 96, 100, 114, *plate 8*
angiosperms, 47, 116, 155–56
anoxia. *See* ocean anoxia
archaeocyathids, 166
Arrow Rocks, New Zealand, 43
atmosphere: GEOCLIM modeling program, 168
atmospheric change: Permo-Triassic extinction, 61–63

bacteria, green sulfur, 44
basaltic magma, 31, 34, *plate 3*
belemnites, 140, 142, 146–48, 152, *plate 15*
bellerophontids, 102
Benedictine abbey (Whitby), 139
bennettitaleans, 94, 108–9
Benton, Mike, xii, 87, 113, 140

Bergman's rule, 70, 98, 162
bicarbonate, 133, 164, 168, 169
biological pump, 81
birds, 155
bivalves: Capitanian crisis, 13, 37, 40, 51; carbon isotope, 126; extinctions of, 126, 134, 158; Lilstock Formation, 127; Permo-Triassic extinction, 52, 86; salinities and, 125; shell of, 84; Smithian/Spathian extinction, 102; Toarcian Stage, 139–43; Triassic Period, 90, 92, 105–6, 128, 137, *plate 7*
black shale: anoxic, 96, 143, 148; of Blue Lias, 127–28; of Cambrian Period, 166; as evidence of gas eruptions, 150–52, 156; Japanese, 144; Jet Rock, 140–41, 146–47, *plate 14*; Panthalassa Ocean, 146; Permian, 150; Selli Event, 158; Sulphur Band, 141; Triassic Period, *plate 6*; transition from, 134; of Westbury Formation; 125
Blue Lias, 125–28
Bonarelli Event, 158
Bond, Dave, xii, xiii, 21, 36, 57–58, *plate 2*, *plate 4*
Bottjer, Dave, xii, 90
Bottrell, Simon, *plate 3*
bounce back, 40, 92, 102, 127, 144
brachiopods: Capitanian crisis, 13, 15–16, 23, 36–37, 40, *plates 1–2*; end of the Triassic, 125, 142; Permo-Triassic extinction, 51–52, 54, 82–83, 86; Tethyan Ocean, 147; Toarcian extinction, 142

breccias: volcanic, 29, 30, 150, 177n2
Bryan, Scott, 177n2
bryozoans, 40, 86, 113, 125
buffering, 37
burrow makers: Permo-Triassic mass
  extinction, 178n11

calcareous algae, 53, 85
calcium carbonate, 59, 82, 84, 115, 164
Cambrian Period, 15, 42, 91, 165–67;
  extinction rates, 15
Capitanian extinction, 4, 5, 12,
  12–38, 25; acidification, 36–38;
  dinocephalians, 13, 27–28, 37–38, 44,
  176; discovery of, 17–18; Emeishan
  volcanism, 18, 26, 29; fusulinacean
  genera, 15–17; limestones, 21, 23; terrestrial vegetation, 27–29; volcanism
  and atmosphere, 35–36
carbonates, 91, 111, 115, 178n16; baking
  rocks of, 34; bicarbonate, 133, 164,
  168, 169; calcium, 59, 82, 84, 164; carbon type, 61; deep-sea, 164–65, 169,
  171, 173; deep-sea mud, 144, 163;
  formation of, 144, 164; lava flows, 19,
  22; microbial, 100; in skeletons, 85,
  105, 116, 168
carbon dioxide: atmospheric concentration, 61, 67, 84; carbon isotope
  ratios, 62–63; carbon pump, 171;
  fossil soil horizons, 133, 178n16;
  generation of, 10, 34–35, 64, 68–69,
  132, 164–65; hypercapnia, 85; ocean
  acidification by, 11, 104; in photosynthesis, 115, 124; removal of,
  73, 168–70, 171, 172, 175; stomata
  density, 124–25; volcanic release
  of, 82–83, 150, 151–52; weathering
  volcanic rock, 133, 160
Carboniferous-Middle Permian interval, 167
Carboniferous Period, 1, 15, 25, 94–95,
  121, 172

carbon isotopes, 178n16; Cambrian
  Period, 166; Carnian Pluvial Event,
  112; end-Triassic extinction, 132;
  Permo-Triassic mass extinction, 53,
  61–63, 68, 72–74; Toarcian crisis,
  146–47, 149–51, 158, 161; Triassic-
  Jurassic boundary, 126–27
Carnian Pluvial Event, 109–16
Carnian Stage, 4, 5, 109, 120; climate
  episode, 117; crisis, 158; Triassic, 5
Carrano, Matthew, 145
Catastrophes and Lesser Calamities
  (Hallam), 128
catastrophism, xv–xvi, 177n5
cause-and-effect scenarios, 10, 29, 73, 87,
  112, 151
$^{13}C/^{12}C$ ratios, 61–62, 68, 103, 177n7
Central Atlantic Magmatic Province
  (CAMP) eruptions, 118–19, 123, 129,
  132–33, 135–36, 138, 149, 154, 157
charcoal, 96, 123
cheirolepidiaceans, 122
chert: cliff-forming, *plate 6*; deep-sea,
  74, 132; radiolarian, 42–43, 59, 115,
  143; red siliceous, 129; Toarcian crisis,
  143–44. *See also* radiolarian chert
Chicxulub: crater at, xvi, 135–36; meteorite impact at, 159–60
China, *plates 3–4, plate 9*
chiniquodontids, 113
chlorine, 31
chloroplasts, 116
chondrichthyians, 91
Christie, Agatha, 87
cicadas, 45, 63
Clapham, Matthew, xii, 36, 37
Cliffhanger (movie), 112
climate belts, 10–11
climate change, 10, 33; linking anoxia
  to, 68; mass extinction and, 113,
  159–60, 175. *See also* global warming
club mosses, 12
coal formation, 47, 49, 54, 74, 94, 101, 172

"coal gap", 100
coccolithophores, 144, 153
coccolithophorids, 115–16, 163, 165
coccoliths, 144–45, 152–53, 155, 163–65, 169–71
cockroaches, 45, 63
coelacanth, 91
Coleoptera (beetles), 95
computer models, 80–81
conodonts: Capitanian crisis, 20–21, 23; extinctions of, 103, 113, 130, 146; microfossil, *plate 9*; oxygen isotope values, 63, 97, 103; Permo-Triassic extinction, 40–41, 52, 86; Smithian/Spathian extinction, 103–5; temperature and, 98–99
contact metamorphism, 34, 65
corals, 15, 53, 93, 106, 138
corystosperm, 55
Courtillot, Vincent, 18
Cretaceous Period, 9–10; 15; bird diversity of, 155; dinosaurs of, 119–20; Early, 154–56, 168; insects in, 155–56; LIPs and life during, 157–65; volcanism in, 156–57
Cretaceous–Tertiary mass extinction, xvi, 3, 159–61
crinoids: Capitanian crisis, 13–15, 36, 40; extinctions of, 142; Triassic period, 110–11, 113
crurotarsans, 114–15, 117, 120–21, 132, 156, 176
crustaceans, 54, 86, 106, 142, 156
cyanobacteria, 43–44, 69, 77–78, 81, 90, 100
cycads, 94
cynodonts, 96

Dactylioceras tenuicostatum Zone, 130–40, 148, 152
Dal Corso, Jacopo, xii, 112
death-by-acidification, 36–37
Deccan Traps of India, 10, 30, 159

denaturation, proteins, 98
deoxygenation: ocean, 55, 63, 75–81, 168
Devonian Period, 1–2, 15, 167
diatoms, 170
dibenzofuran, 48
dicerocarditids, 127
Dicroidium, 47, 55, 109
dicynodonts, 27, 38, 44, 94–95, 108, 113
Dilophosaurus, 145
dinocephalians: Capitanian extinction, 13, 27–28, 37–38; Permo-Triassic extinction, 44, 176
dinoflagellates, 115
dinosaurs, 2, 8, 116, 117, 176; bipedal, 108; Chicxulub crater, 135–36; Cretaceous Period, 119–20, 155, 160–61; death of, xv, 10, 92; first primitive, 114–15; Jurassic Period, 119–20, 145; mass extinction of, 2–3, 6, 120, 135–36, 159; Triassic life, 121
Diptera (true flies), 95
dolomites, 34
Dolomites Mountains: Italy, 50–51, 90, 112, *plate 11*
Donnadieu, Yannick, 168
downwelling, 79

Earlandia foram, 23–24, 51
Earth: resilient planet, 167–76
Earth system models of intermediate complexity (EMICs), 80
echinoids, 40, 106
Emeishan Traps, 17, 31, 36, 39, 64, 113, 149
Emeishan volcanism, 18, 22–23, 26, 29, 31–32, 177n2
end-Permian mass extinction. *See* Permo-Triassic mass extinction
end-Triassic mass extinction, 6; acidification and anoxia of, 134–35; carbon isotope spike in, 132; kill mechanisms of, 132–33; losses in seas and

end-Triassic mass extinction (*continued*)
oceans, 125–28, 130; marine, 127;
recovery from, 137–38; reef, *plate 10*;
role of meteorite impact, 135–36;
timing of, 131–36; victims of, 119–30
Erwin, Doug, 87
euxinic environment, 58–59, 177n6
Evolutionary Catastrophes (Courtillot), 18
extinction model, 68–75

Faroe Island, *plate 16*
fern spike, 122–23, 127, 131, 135
Ferrar, Hartley, 138
fish: Early Triassic, 99–100; Permo-
Triassic extinction, 41
fluorine, 31
foraminifers, 14
forams, 14, 54; agglutinating, 83; Capitanian crisis, 40; *Earlandia*, 23–24,
51; extinctions of, 142; fusulinaceans,
15–16, 23, 36; Permo-Triassic extinction, 51, 162; planktonic, 155, 159
Forbrydelsen (Danish TV show), 88
fossil soil horizons, 133, 178n16
framboids: pyrite, 56–59, 144
Frasnian and Famennian Stages boundary (F-F event), 167
Fröbisch, Jörg, 107
Fröbisch, Nadia, 107
frogs, 45
fungal spores, 48
fusulinaceans, 14–15, 23, 36

Galfetti, Thomas, 104
gastropods, 13, 40, 86, 90, 106, 137, 142
GEOCLIM modeling program, 168
gigantopterids, 12, 47
Ginkgo biloba, 108
global environment: Early Triassic,
96–102; GEOCLIM modeling
program, 168
global warming, 10; Carnian crisis,
158; greenhouse gases, 147; ocean

deoxygenation, 78–79, 175; Permo-
Triassic mass extinction, 69, 72–73,
75; PETM (Paleocene-Eocene thermal maximum), 161, 163; resilient
planet, 168; Toarcian Age, 151. *See
also* climate change
glossopterids, 12, 47
Gondwana, 6–8, 94, 108, 153, 154, 172
gorgonopsids, 27, 44, 45
greenhouse: Early Triassic, 96–102;
global warming, 147
greenhouse gases, 35–36, 60, 68, 80, 132,
147, 156, 162, 168
Greenland, 46, 49, 54, 58, 122, 161
green sulfur bacteria, 44, 56
Gulliver's Travels (Swift), 90
gymnosperms, 12–13, 25, 46–47, 54, 103

habitat destruction, 10
Hallam, Tony, xi, 50, 56, 110, 128, 139,
178n19
halogens, 31–32, 35, 38, 69, 71, 73, 135,
174
Hauterivian stage, 154–55
Hautmann, Michael, 134
Hawaii, 10, 31, 133
Heteroptera (true bugs), 95
Heydari, Ezat, 82
Hilton, Jason, xii, 21, 25, 26
holocephalians, 14
holothurians, 86
hopanes, 43–44
hydrogen chloride (HCl), 67
hydrogen sulfide, 44, 57, 82–83, 144, 174
hydroxyl radical, 72
Hymenoptera (wasps and bees), 95
hypercapnia, 85–88

ichthyosaurs, 106–7, 130, 132, 137
Ichthyosaurus, 130
Inaperturopollenites, 46, 54
insects: Permo-Triassic extinction, 45;
Triassic, 95–96, 117

Isoetes, 48
isorenieratane, 44
Isozaki, Yukio, xii, 18, 42

Jerram, Dougal, xii, 30, *plate 4*
Jet Rock, 140, 141, 144, 146–47, 178n17, *plate 14*
Jiang, Haishui, xii, 20, 23
Jin, Yugan, 14
Jinogondolella species, 23, 177n1
Joachimski, Michael, 63, 97
Junggar Basin China, 123
Jurassic: Early, 8
Jurassic Park (movie), 145
Jurassic Period: dinosaurs, 119–20; extinction rates, 15; seafloor life, 137; subdivisions of geological time, 4; Toarcian Stage, 6, 139–40

Kalkarindji Province eruptions, 166
Karoo Basin, 138, 150
Karoo-Ferrar eruptions, 138, 141, 149, 151–53, 154
Khuff Formation, 60
Kiaman Reverse Superchron, 19, 23, 26, 28
Knoll, Andy, 85, 87–88

Lai, Xulong, xii, 20, 23
Langport Member, 134
Laurasia, 6, 7
Lilliput effect, 90, 178n9
Lilstock Formation, 125–27
limestones, 53, 117n3; Alpine, 127–28; Carboniferous, 21; carbon isotope, 61–62, 68, 103; deposition, 23, 112, 168–69; Emeishan lavas, 34; formation of, 59, 111, 164, 171–72; Khuff Formation, 60; Lilstock Formation, 125–27; Maokou Limestone, 19–22, 30; marine, 127, 132, 134; ocean acidification, 82; pelagic, 163–65; Rhaetian Sea, 125; South China, 97;

Tethys, 155; Tibet, 143; Triassic, 60, 90, 111, *plate 5*; Xiong Jia Chang, 22–23; Yunnan Province, 29
Lingula, 51
LIPs (large igneous provinces), 30; CAMP (Central Atlantic Magmatic Province), 118–19, 132, 138; conference talks, 9; Deccan Traps, 159; Emeishan LIP, 17–18, 22, 24; environmental change, 168; eruptions, 33–35, 37, 67, 165–67, 173, 175; Frasnian and Famennian Stages (F-F event), 167; Kalkarindji Province, 166; Karoo-Ferrar eruptions, 149, 152; life after Jurassic Period, 157–65; NAIP (North Atlantic Igneous Province), 33, 157, 161, 173–74; of Pangea, 17, 153, 161, 165; Paraná-Etendeka eruption, 154, 156–57; volcanic regions, 9–10
lissamphibians, 45
lithiotids, 143
Little, Crispin, 140
Looy, Cindy, xiii, 46, 54
Lueckisporites, 54
Lunatisporites, 54
Lundbladispora, 54
lycopsids, 12, 25, 46–47, 54–55, 94, 103
Lystrosaurus, 94–95, 103, 108

McArthur, John, xiii, 147
McElwain, Jenny, xiii, 124, 134, 148
Mackenzie, Clara, 178n12
magnetostratigraphy, 22, 23, 26, 28
mammals: island colonization, 121–22
mantle plume, 67
Maokou Formation, 19
Maokou Limestone, 19–22, 30
Marzoli, Andrea, 118, 132
mass extinction: Cretaceous-Tertiary crisis, xvi, 3, 159–61; definition, 2–3; Pangea + LIP volcanism, 9–10. *See also* end-Triassic mass extinction;

mass extinction (*continued*)
Permo-Triassic mass extinction;
Toarcian crisis
Mass Extinctions and Their Aftermath
(Hallam and Wignall), 178n19
Mattioli, Emanuela, 152–53
megalodontids, 105, 127, 143, 156, 158,
plate 10
megalosaurids, 145
Mesozoic Era, 3
Metacopina, 142–43
meteorite impact, xv–xvi, 167;
Cretaceous-Tertiary mass extinction,
159–61; end-Triassic crisis, 135–36;
Yucatan Peninsula in Mexico, 3, 6, 10
methane, 62, 63; generation of, 34,
72–73; release from hydrates, 69,
72–73, 82, 132, 150–51
Meyer, Katja, 81
Middle Permian times, 8
model: extinction, 68–75
Mount Pinatubo eruption, 9, 32
Murder on the Orient Express (Christie), 87
'Murder on the Orient Express'
scenario, 87, 88, 104

Nanjing Institute of Geology and Palae-
ontology (NIGP), 14, 21, 51
nannoplankton, 116
nautiloids, 14, 40, 86
neopterygians, 91
Neospathadus, plate 9
Newark Basin, 131
Newell, Andrew, 87
Newton, Rob, xii, xiii, 21, 58, *plate 12*
nitrogen isotope ratios, 43
Norian Stage, 4, 120–21, 130, 136
North Atlantic Igneous Province
(NAIP), 33, 157, 161, 173–74
Norway, *plates 1–2, plate 6*

ocean acidification, 82–85, 83, 104, 116,
134, 152, 163–65, 173

ocean anoxia, 158, 173; Bonarelli Event,
158; global warming and, 60, 88, 151;
Permo-Triassic extinction, 69, 75, 77,
80–81, 134, 148–49, 151, 153; Triassic-
Jurassic boundary, 129
ocean circulation, 77–79
ocean deoxygenation, 55, 63, 75–81, 168
Ontong Java Plateau, 158
Ordovician Period, 2, 15
osmium, 130, 131
ostracods, 40, 52, 54, 86, 125, 142–43, 162
overhunting, 10
oxyconforming, 99
oxygen isotopic ratio, 36, 63, 68, 97–98,
103, 147–48
oxygen-minimum zone (OMZ), 76
ozone destruction, 38, 69, 71, 73–74, 88,
135, 174
ozone formation, 72, 174

Paleocene-Eocene thermal maximum
(PETM), 161–63
paleomagicians, 19
paleomagnetism, 19
Paleozoic Era, 3, 142
Pangea, 1–3; breakup of, 110, 118, 153–
54, 171, 173; eruption at time of, *plate
16*; existence of, 8–9; single continent
of, 5, 7; subdivisions of geological
time during time of, 4; weathering,
168–69, 172; world map of, 118
Pangean crises: hypotheses for, 11
Panthalassa Ocean, 1, 7, 8, 113; black
shales of, 146; Carnian carbonates of,
115; map of, 118; ocean-floor sediment, 112, 157; radiolarian extinction,
129, 143–44; species loss, 39; surface
waters of, 125; ventilation of, 80
Paraná-Etendeka eruptions, 154, 156–57
Paton, M. T., 178n14
Payne, Jonathan, xiii, 36, 37
Permian Period, xv, 1, 3, 5, 7; brachiopods, *plate 1*; carbon isotope changes

during, 53; carbon isotope ratio of limestone, 62–63; end of the, 39, 40–44, 40–47, 59, 64, 68; extinction rates in, 15; flora of, 25, 55; fossils of, 51–52; Late, 27, 38, 43, 47, 60; Middle, 23, 24, 25, 36; species of, 54; subdivisions of geological time, 4; world map of, 7

Permo-Jurassic period, 3, 10

Permo-Triassic mass extinction, 3–5, 39, 40–49, 162; acidification role in, 84; aftermath of, 89; atmospheric change in, 61–63; burrow makers in, 178n11; carbon isotope ratios, 53, 61–63, 68, 72–74; carbon isotope spikes in, 73, 103, 112, 127; crinoids in, 111; deoxygenation of ocean, 75–81; dicynodonts in, 113; double-punch of, 166; echinoderms in, 110; flow chart for events of, 69; "Forbrydelsen" approach to, 88; fossil record in, 138; generic extinction percentages, 86, 178n13; global warming in, 163, 173; insects in, 95; killing seas of, 55–60, 82; LIP volcanism in, 119; meandering rivers of, 26, 28; model of, 68–75; ocean anoxia during, 69, 75, 77, 80–81, 134, 148–49, 151, 153; ozone damage in, 71, 72; parallels to end-Triassic extinction, 127, 129–30, 142; plants in, 25, 45–49, 54–55, 122; radiolarians of, 144; scenario for, 87; Siberian Traps, 64–68; survivors of, 91–96, 97, 108; timing of, 50–55; Toarcian crisis similarity to, 149; volcanism and, 17–18

pH: calculation of, 177n3; ocean, 37, 83, 84, 153, 169

photosynthesis: at high temperatures, 100–101, 173; carbon dioxide for, 115, 124, 151, 163, 165; organic matter formation, 61, 170; phytoplankton, 43–44, 76

phreatomagmatic eruption, 30–32

phytoplankton, 43–44, 69, 76, 100, 151

phytosaurs, 114, 120, 121

pillow lavas, 31

placodonts, 106

planktonic groups, 42, 115, 173

plants: atmospheric carbon dioxide, 170–71; fossils as clues to atmospheric change, 124–25; Permo-Triassic crisis, 25, 45–49, 54–55, 122

Pleuromeia, 103, 108

Pliensbachian Stage, 4, 140, 142

Pliensbachian-Toarcian boundary, 141, 143, 148, 151

plumbing system, 34–35, 119, 166

pollen grains, 46–48, 122–23, 127

polychaete worms, 86

polycyclic aromatic hydrocarbons (PAHs), 123

Pontus Euxinus, 177n6

post-eruption creatures, 23

Precambrian oceans, 91

Preto, Nereo, xiii, 112

primates, 162

Psiloceras, 128

pteridosperms (seed ferns), 12, 54, 94

pterosaurs, 114, 116, 117, 120

pyrite: framboids, 56–59, 144

radiolarians: Carnian crisis, 115; chert, 42–43, 59, 115, 143; end-Triassic crisis, 125, 129–30, 134; Permo-Triassic extinction, 42–43, 53, 59, 68–70, 77–78, 83, 86; planktonic, 143; Toarcian crisis, 144–45

rauisuchians, 114, 120

Rebellatrix, 91

red algae, 116

redlichiids, 166

Reduviasporonites, 48

reefs: construction of, 84, 94, 166; coral, 14, 106, 117, *plate 10*; disappearance of, 5, 68, 127; recovery of, 109, 138; Tethyan, 125

reptiles, 27, 122; Carnian extinctions, 114; marine, 99, 100, 106, 130, 137, 140, 159; pareiasaurs, 44
Retallack, Greg, 28–29, 97, 117n4
Rhaetian Stage, 4, 120–22, 125, 127–28, 130, 132, 134–36
rhynchosaurs, 113
roadcut, 22, 30
Rochechouart impact: crater, 135–36
rock weathering, 110, 164, 171–73
rudists, 158
Ruffell, Alastair, xiii, 110–12
rugose corals, 15, 86
Rumsfeld, Donald, 33
Russell, Joellen, 178n10

Sanmiguelia, 114
Saurichthys, 91–92
Schindewolf, Otto, 177n5
scleractinian corals, 94, 106, 127
Scythian, 89–90
sea surface, 174; evaporation of, 79; temperature of, 63, 68, 80, 97–98, 103
seawater: carbon ratios of, 61; isotope ratios of, 146, 151; osmium content, 131; risk of reaching molten rock, 30; temperature of, 98, 104, 147–48; uranium in, 59
seismites, 126, 135–36
Selli Event, 158
Sephton, Mark, xiii, 48
Shastasaurus, 107
Shcherbakov, Dmitry, 45
Shen, Shu-zheng, 51
Shihhotse Formation, 25–26
Shonisaurus, 107
Siberian Traps, 7, 39, 64–68, 73, 87, 96, 113, 119, 132, 135, 149, 167, 178n14
Sichuan Province, 21
Simms, Mike, xiii, 110–12, 113
Smithian/Spathian crisis, 5, 102–5; carbon and oxygen isotope changes, 103; cause and effect in, 104, 112–13; extinction rate of, 15

Sobolev, Alexander, 66, 67, 71
Sobolev, Stephen, 66, 67, 71
Song, Haijun, xii, 52
sphenodontians, 121
Spitsbergen, 36, 58, plates 1,2 & 6
sporomorphs, 122–23, 133
spumellarians, 143–44
stairs, 9
Stallone, Sylvester, 112
Stanley, Steve, 14, 17
Stevens, Liadan, 25, 26
stromatolites, 90–91
sulfur dioxide (SO$_2$), 31–32, 35, 64, 68, 135
Sulphur Band, 141
Sun: ultraviolet radiation, 72
Sun, Yadong, xii, 20, 23, 97, 99, plate 4, plate 9
superchron, 19, 23, 26, 28
supercontinent. See Pangea
Surtsey, Iceland, 31
Svensen, Henrik, 33, 34, 65, 133, 150, 156, 161
Svensen's hypothesis, 34, 65, 133, 150
Swift, Jonathon, 90

temnospondyls, 44–45, 95, plate 8
Tethyan Ocean, 7, 8, 100, 139, 141–42, 147
Tethys, 8, 48, 70, 74, 92, 117, 128, 142–44, 152
tetrads, 46
Thalattoarchon, 107
thalattosaurs, 106, 113
thrombolites, 90–91
Tibet, 59, 128
timing: Permo-Triassic mass extinction, 50–55
titanopterids, 118
Toarcian crisis, 142–48, 158, 163; geochemical evidence in belemnites, 146–47; Jet Rock, plate 14; lithiotids, 143; planktonic radiolarians, 143–45; temperature trends, 148–53

Toarcian Stage: Jurassic, 6, 139–40
Toba eruption, 33
Toggweiler, Robert, 178n10
Tong, Jinnan, 52
Towapteria, 51
traps, 9; Deccan Traps, 10, 30, 159;
    Emeishan Traps, 17, 31, 36, 39, 64,
    113, 149; Siberian Traps, 7, 39, 64–68,
    73, 87, 96, 113, 119, 132, 135, 149, 167,
    178n14
Triassic-Jurassic boundary: warming-
    cooling oscillations, 133; world map,
    118
Triassic Period, 3; carbon isotope
    changes during, 53; Carnian Pluvial
    Event, 109–16; Carnian Stage of,
    5–6, 109; end of, 8; extinction rates,
    15; flora, 55; greenhouse in Early,
    96–102; insects in, 95–96; sandstone
    from Late, *plate 13*; Scythian (Early
    Triassic), 89–91; Smithian/Spathian
    extinction in Early, 102–5; smooth
    sailing in Late, 104–9; subdivisions of
    geological time, 4; survivors of Early,
    91–96; world map of Early, *plate 8*.
    See also end-Triassic mass extinction;
    Permo-Triassic mass extinction
turtles, 106
Twitchett, Richard, xii, 46, 90

Ukstins-Peate, Ingrid, 177n2
upwelling, 82
Ural Mountains, 7, 27, 45
uranium, 59–60, 131
Utatsaurus, 107
UV radiation, 71

venerid bivalves, 106
violent shocks, 10
Visscher, Henk, xiii, 46, 48, 71

volcanism, 3, 5, 10; climate conse-
    quences of, 36; Emeishan, 18, 22–23,
    26, 29, 31–32, 177n2; giant, xvi, 10, 17,
    22, 167; LIP, 9; Pangea, 8–9; Permo-
    Triassic mass extinction and, 17–18

wallowaconchids, 127
Wang, Wei, xiii, 21
Wangpo Bed, 21
Ward, Peter, 49
weathering, rocks, 172
Westbury Formation, 125, 126
West Siberian Basin, 39, 64
Weylandites, 54
Whitby (England): ammonites in,
    139–41; belemnites in, 147; Jet Rock,
    140, 141, 144, 147, 178n17, *plate 14*
Widdowson, Mike, *xiii*, 30, *plate 4*
Winguth, Arne, 80
Winguth, Cornelia, 80
Wordian Stage, 4, 28
world map, 7, 12, 118, *plate 8*
worms, 83, 86, 178n11
Wrangelia 113

Xiong Jia Chang: rocks of, 22, 23

Yadong, Sun, 20, 23, 97, 99, *plate 9*
Yang Xiangning, 14, 17
Yin, Hongfu, xii, 52
Yucatan Peninsula: meteorite impact,
    xvi, 3, 6
Yunnan Province, 29

Zachos, James, 162
zircon, 131, 178n14
zone fossils, 41
Zoophycus, 86
zooplankton, 76–77, 81
Zygnematales, 48